AF752263

Dielectric Resonator Antennas

For a complete listing of titles in the
Artech House Antennas and Propagation Library,
turn to the back of this book.

Dielectric Resonator Antennas

Biswajeet Mukherjee

Monika Chauhan

ARTECH
HOUSE

BOSTON | LONDON
artechhouse.com

Library of Congress Cataloging-in-Publication Data
A catalog record for this book is available from the U.S. Library of Congress.

British Library Cataloguing in Publication Data
A catalogue record for this book is available from the British Library.

Cover design by Charlene Stevens

ISBN 13: 978-1-63081-870-8

© 2022 ARTECH HOUSE
685 Canton Street
Norwood, MA 02062

All rights reserved. Printed and bound in the United States of America. No part of this book may be reproduced or utilized in any form or by any means, electronic or mechanical, including photocopying, recording, or by any information storage and retrieval system, without permission in writing from the publisher.

All terms mentioned in this book that are known to be trademarks or service marks have been appropriately capitalized. Artech House cannot attest to the accuracy of this information. Use of a term in this book should not be regarded as affecting the validity of any trademark or service mark.

10 9 8 7 6 5 4 3 2 1

Contents

Acknowledgments

Successful projects rely on the constant support and motivation of one's personal network, as our experience with this book has underlined. Accordingly, we would like to take this opportunity to extend our sincere gratitude to our families. In particular, we thank Papya Mukherjee, who is always a key source of inspiration, and Jayashree Mukherjee for her tenacity. Similarly, T. R. and Usha Chauhan and Navin Kumar have been instrumental in our completion of this work, helping us to sustain our perseverance.

We also thank the student group of Sanghmitra, Daniel Akhil, and Anil Rajput for their constant help and support and offer a special tribute to Professor Jayanta Mukherjee (Department of Electrical Engineering, IIT Bombay) for his blessing of this project. We dedicate the book to the memory of the late Abhoy Kumar Mukherjee.

Last but not least, we thank the almighty, who has helped transform every difficulty to an opportunity.

1

Introduction to Dielectric Resonator Antennas

The necessity for high speed and a high data rate of communication has escalated the frequency of operation of dielectric resonator antennas (DRAs). Communication technology has led to significant improvements in antenna research for designing low-loss, wideband, and high-gain antennas. Significant work in this regard is being done in the field of planar antenna counterparts. However, a higher frequency of operation raises concerns about the nonlinear ohmic losses at these antennas' high frequency of operation.

DRAs offer no ohmic or conductor loss due to the absence of conducting materials, thus making them high-efficiency radiators. In addition, the sudden transition of the dielectric air interface has led to a significant improvement in their bandwidth of operation. Due to these advantages, DRAs offer competitive prospective commercial applications. This book discusses DRAs and their applications in various industrial, commercial, and research sectors.

1.1 Genesis of Dielectric Materials as Radiators

Professors Rajeshwari Chatterjee and R. N. Chatterjee at the Indian Institute of Science (IISc) Bengaluru, India, have contributed significantly to the exploration of dielectric materials as potential radiators [1]. Their research group explored the loading of the dielectric materials on the complete aperture of horn antennas. Their detailed investigation covered all forms of the horn antennas with rectangular and cylindrical cross-sections. The loading of the dielectric on an aperture led to interesting findings in terms of near-field and far-field analysis.

Since the naturally occurring dielectric materials have a higher permittivity, dielectric resonators are well-known for their use in oscillator and filter applications. A higher permittivity means a higher quality factor, and so the dielectric resonator can act as a good storage element. In 1983, S. A. Long and his research group first identified the potential of a dielectric material capable of radiating as well [2]. In their investigation, a rectangular block of the DRA, popularly known as the rectangular DRA (RDRA), was placed on a ground plane and suitably excited by a coaxial probe feed. This research was then extended to the other basic geometries, but confined to the three coordinate systems. The three basic geometries include the RDRA for rectilinear coordinate systems, the cylindrical DRA (CDRA) [3] for cylindrical coordinate systems, and the hemispherical DRA (HDRA) [4] for spherical coordinate systems.

Full spheres are not used as radiators, since they have only one point of contact on the ground plane and are, thus, mechanically unstable for practical or commercial applications. All three geometries have their own flexibilities and capabilities. The difference in the basic shapes also leads to a comprehensive study of different resonant modes excited under different feeding conditions.

There are two different types of modes excited in a DRA: confined and nonconfined modes. The modes define the behavior of the electric field (E field) and the magnetic field (H field) in the near-field and far-field regions. The electromagnetic mode nomenclature adheres to the geometry and the coordinate axis under consideration.

Any resonant mode should satisfy two important conditions of the E field and H field, as in the following:

$$\vec{E} \bullet \hat{n} = 0 \tag{1.1}$$

$$\hat{n} \times \vec{H} = 0 \tag{1.2}$$

where $\hat{n}$ is the normal-to-surface of the resonator, $\vec{H}$ is the M field intensity, and $\vec{E}$ is the E field intensity. The modes that satisfy the conditions in (1.1) and (1.2) are called *confined modes,* whereas those that satisfy only (1.1) are called *nonconfined modes* [5]. As shown by Van Bladel in [6], confined modes can be satisfied only by the geometries that support the body of revolution (i.e., cylindrical and hemispherical geometries), whereas the nonconfined modes are satisfied only by rectangular geometries, which do not support the body of revolution. A detailed description of the modes can be found in [7].

While discussing the fields, two important points should be taken into account; the first is coupling. The coupling current can be either modeled as an electric vector current distribution ($\vec{J}_S$) or a magnetic current vector

$\overrightarrow{M_S}$. Let the amount of coupling be given by γ, then the fields within the DRA can be determined by applying the Lorentz reciprocity theorem with an appropriate boundary condition for the specified shape of DRA. Thus, for $\vec{J}_S$,

$$\gamma \propto \int \left(\vec{E}_{DRA} * \vec{J}_S \right) dV \quad (1.3)$$

and for $\overrightarrow{M_S}$,

$$\gamma \propto \int \left(\vec{H}_{DRA} * \overrightarrow{M_S} \right) dV \quad (1.4)$$

where V is the volume occupied by the source within which the current vectors (electric or magnetic) exist, while the field vectors $\vec{E}_{DRA}$ and $\vec{H}_{DRA}$ exist within the DRA. This means that for efficient coupling, the source current—whether through probe, slot, or microstrip—should be located within the areas of strong field configuration in the DRAs and hence, the different mode nomenclatures exist in DRAs. For electric field coupling, the electric current should be colocated with the electric fields and for strong magnetic coupling, the magnetic vector current should be colocated with the magnetic field maxima areas.

In examining resonator theory, we learn that the DRAs have natural resonant frequencies dependent on the resonant dimensions, which are calculated based on the modes, and this needs to be loaded with the impressed frequency from the external fields or excitation. This phenomenon affects the quality factor (Q factor) of the DRA. The external Q factor (Q_{ext}) can be defined in terms of coupling factor γ as

$$Q_{ext} = \frac{Q}{\gamma} \quad (1.5)$$

Where Q is the unloaded Q-factor and the loaded Q factor (Q_l) is expressed as

$$Q_l = \left(\frac{1}{Q} + \frac{1}{Q_{ext}} \right)^{-1} = \frac{Q}{1+\gamma} \quad (1.6)$$

Maximum power is transferred for $\gamma = 1$; $\gamma < 1$ is the under-coupled, and $\gamma > 1$ is the over-coupled case of DRA. The Q factor of any DRA is estimated as follows [7]

$$Q = 2\omega_0 \frac{\text{stored energy}}{\text{radiated power}} \propto 2\omega_0 \left(\varepsilon_r \right)^p \left(\frac{\text{volume}}{\text{surface area}} \right)^s, \text{where } p > s \geq 1 \quad (1.7)$$

Thus, as the Q factor decreases, the bandwidth of operation increases, which often raises the question: Why does a DRA radiate, despite consisting of considerably high-permittivity materials? In fact, the presence of high-permittivity material shows that the Q factor is high and so the DRA radiation mechanism and dielectric materials act as a storage element, as has been employed in ceramic filters or oscillators.

However, the relation in (1.7) simplifies the understanding: For the fields to radiate rather than store, the surface area-to-volume ratio of the DRA needs to be increased, thus reducing the Q factor. This also aids in improving the radiated power-to-stored energy ratio.

Further perturbation theory has been extensively employed to reduce the effective dielectric constant, which also aids in improving the radiation characteristics, the impedance bandwidth of operation, and with the gain and directivity of the DRA.

1.2 RDRAs and Their Modes

RDRA supports TE_{mnp} and TM_{mnp}, where m, n, p represents the half-cycle sinusoidal variations of the transverse field component along the x, y, z axes, respectively, of the rectilinear coordinate system. In [8], the H modes are referred to as transverse magnetic (TM) modes, and E modes are referred to as transverse electric (TE) modes. It is observed that equation (1.1) is not satisfied by TM_{mnp} modes by using the dielectric waveguide model (DWM) after applying the magnetic wall condition, so the TM modes are not observed in RDRA. By using the DWM and presuming z to be the direction of propagation or the current vector direction, the fields in RDRA [5, 9–10] are predicted as follows:

$$H_x = \frac{(k_x k_z)}{j\omega\mu_o}\begin{Bmatrix} sin(k_x x)\cos(k_y y) \\ cos(k_x x)\cos(k_y y) \end{Bmatrix}\sin(k_z z) \tag{1.8}$$

$$H_y = \frac{(k_y k_z)}{j\omega\mu_o}\begin{Bmatrix} \cos(k_x x)\sin(k_y y) \\ \sin(k_x x)\sin(k_y y) \end{Bmatrix}\sin(k_z z) \tag{1.9}$$

$$H_z = \frac{(k_x^2 + k_y^2)}{j\omega\mu_o}\begin{Bmatrix} \cos(k_x x)\cos(k_y y) \\ \sin(k_x x)\cos(k_y y) \end{Bmatrix}\cos(k_z z) \tag{1.10}$$

$$E_x = k_y \begin{Bmatrix} \cos(k_x x)\sin(k_y y) \\ \sin(k_x x)\sin(k_y y) \end{Bmatrix} \cos(k_z z) \tag{1.11}$$

$$E_y = -k_x \begin{Bmatrix} \sin(k_x x)\cos(k_y y) \\ \cos(k_x x)\cos(k_y y) \end{Bmatrix} \cos(k_z z) \tag{1.12}$$

$$E_z = 0 \tag{1.13}$$

While, *l*, *w*, and *d* are the length, width, and depth of the RDRA (dimensions), *c* is the velocity of light in free space, and the wave numbers are defined as follows:

$$k_z \tan\left(\frac{k_x d}{2}\right) = \sqrt{(\varepsilon_0 - 1)k_{mn}^2 - k_z^2} \tag{1.14}$$

$$k_{mn} = \frac{2\pi f_m}{c}, k_x = m\frac{\pi}{w}, k_y = n\frac{\pi}{d} \tag{1.15}$$

$$k_c^2 = k_x^2 + k_y^2 = \varepsilon_0 k_{mn}^2 - k_z^2 \tag{1.16}$$

From (1.8) to (1.13), the upper functions in the bracket are chosen when *m* and *n* are both odd, whereas the lower functions are chosen when *m* is even and *n* is odd. It is to be noted that both *m* and *n* cannot be even at the same instance. The modes in RDRA are often represented by $TE^{Z}_{mn\delta}$. Here δ instead of *p* shows that while the direction of propagation is *z*, the half-cycle of sinusoidal rotation is incomplete along the *z*-axis.

The RDRA offers two degrees of freedom, allowing the flexibility to alter the ratio of *w*/*l* and *w*/*d*. The lowest order TE mode is the TE_{111} mode as observed in the RDRA. Figure 1.1 shows the H-field orientation for various TE modes [7, 10]. Figure 1.1(a, b) shows the TE^{x}_{121} and TE^{y}_{211} plots, and Figure 1.1(c, d) shows the TE^{x}_{212} plots.

The design of the RDRA for varied applications and its two degrees of freedom has led to an extensive investigation of the geometry into various modified forms so as to achieve compactness, wideband function, and high-gain

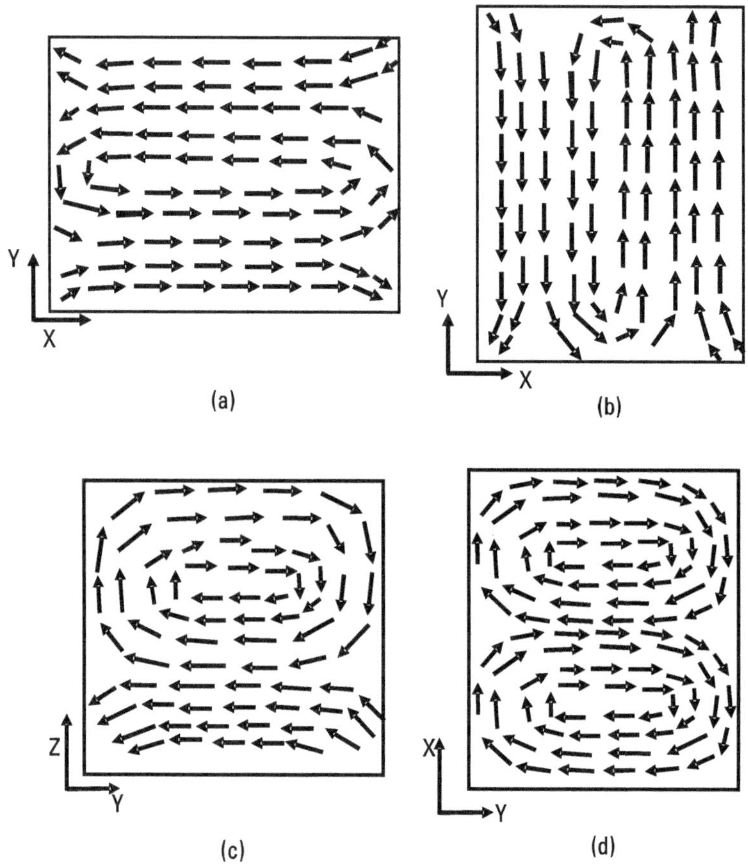

Figure 1.1 The modes of RDRA: (a) TE^{x}_{121} mode, (b) TE^{y}_{211} mode, (c) TE^{y}_{212} mode along the y-z plane, and (d) TE^{y}_{212} mode along the x-y plane.

operations. Different feeding mechanisms have also been explored in the same so as to achieve circular polarization (CP).

It is also interesting to note that while there can be a square cross-section in the rectilinear geometry, the cubical geometry remains a constraint to be explored. This can be explained based on the field equations, (1.8) to (1.13), where *m* and *n*, individually or together, always take alternate odd values and cannot be even numbers. This is explored in [7, 10–13].

In a simple example from [13], a dielectric material of $\varepsilon_r = 10.2$ and a loss tangent of $\tan\delta = 0.0023$, with dimensions of 3.1cm × 3.1cm × 1.55cm resonates at 2.4 GHz based on the DWM model calculations. This excites the fundamental TE_{111} mode for a probe placed an offset of 0.95 cm from the center. The ground plane is chosen to be large enough so as to avoid any fringing field effects.

1.3 CDRA and Its Modes

In tandem to the modes of the RDRA, CDRA modes are half sinusoid variations along the azimuth (Φ), radial (r), and axial (z) axes, respectively. The modes of a CDRA are grouped as TE_{mnp}, TM_{mnp}, and HEM_{mnp}, where m, n, and p represent the half-cycle sinusoidal variation along the azimuth (Φ), radial (r), and axial (z) axes, respectively. In some cases the p is represented by δ to signify an incomplete half-cycle sinusoidal rotation along the axial axis, if z is the direction of propagation or the exciting current. It is interesting to observe that the TE and TM modes are axially symmetric (i.e., the variation is independent of the azimuth plane), whereas the hybrid electromagnetic (HEM) modes show dependence along the azimuth plane [14–16].

The HEM mode is called the HE mode if E_z is dominant, or EH if H_z is dominant [15]. The dominant modes of the CDRA are classified as $TM_{01\delta}$, $TE_{01\delta}$, and $HE_{11\delta}$. For an isolated CDRA, the dominant mode resonant frequencies can be arranged as $TE_{01\delta} < TM_{01\delta} < HE_{11\delta}$.

The TM mode can be excited in CDRA whenever the direction of excitation current coincides with the axis of the CDRA. This can be easily achieved by feeding a probe at the center of the CDRA. To understand this analogy, consider the magnetic field right-hand rule for the direction of the H fields. This leads to a radiation pattern that resembles a short electric monopole that has a null in the broadside direction. Similarly if the probe is at an offset from the center of an isolated CDRA or displaced from the axis of the cylinder, it is possible to excite the $HE_{11\delta}$-dominant mode, which radiates like a short horizontal magnetic dipole. Note that this discussion assumes that the probe or the exciting current is along the z direction.

In case the configuration of the CDRA has an axis parallel to the groundplane example—a half-split CDRA or sectored split CDRA—a slot feed may excite the $TE_{01\delta}$ mode, which radiates like a short magnetic monopole. For the TE mode to be excited, the axis of the CDRA and the direction of current vector should be perpendicular to each other, as is evident from the various CDRA geometries, including the half-split CDRA. This is detailed in [14–16].

The different modes can be excited by various feeding mechanisms and different orientations of the CDRA. The CDRA offers one degree of freedom to an antenna engineer to facilitate the variation in the radius-to-height (r/h) ratio, for obtaining a suitable resonant frequency and bandwidth of operation.

The approximate fields for a $TE_{01\delta}$ mode of CDRA are estimated as follows [15, 17-20]

$$H_Z \propto J_0(\beta r)\cos\left(\frac{\pi z}{2h}\right) \tag{1.17}$$

$$H_r \propto J_1(\beta r)\sin\left(\frac{\pi z}{2h}\right) \tag{1.18}$$

$$E_\phi \propto J_1(\beta r)\cos\left(\frac{\pi z}{2h}\right) \tag{1.19}$$

$$E_Z = E_r = H_\phi = 0 \tag{1.20}$$

where $J_0(\beta r)$ and $J_1(\beta r)$ are the zeroth and first-order Bessel functions of the first kind, respectively. β is the solution to $J_0(\beta a) = 0$, with a as the radius of the CDRA. The $TM_{01\delta}$ are similar to (1.17) to (1.20), with the magnetic and electric field components interchanged.

On a similar note, the field components of the $HE_{11\delta}$ mode of CDRA are estimated as follows [2, 9–11]

$$E_Z \propto J_1(\alpha r)\cos\left(\frac{\pi z}{2h}\right)\begin{Bmatrix}\cos\phi \\ \sin\phi\end{Bmatrix} \tag{1.21}$$

$$E_r \propto \frac{\partial J_1(\alpha r)}{\partial(\alpha r)}\sin\left(\frac{\pi z}{2h}\right)\begin{Bmatrix}\cos\phi \\ \sin\phi\end{Bmatrix} \tag{1.22}$$

$$E_\phi \propto J_1(\alpha r)\sin\left(\frac{\pi z}{2h}\right)\begin{Bmatrix}\sin\phi \\ \cos\phi\end{Bmatrix} \tag{1.23}$$

$$H_r \propto J_1(\alpha r)\cos\left(\frac{\pi z}{2h}\right)\begin{Bmatrix}\sin\phi \\ \cos\phi\end{Bmatrix} \tag{1.24}$$

$$H_\phi \propto \frac{\partial J_1(\alpha r)}{\partial(\alpha r)}\cos\left(\frac{\pi z}{2h}\right)\begin{Bmatrix}\cos\phi \\ \sin\phi\end{Bmatrix} \tag{1.25}$$

$$H_Z \approx 0 \tag{1.26}$$

where α is the solution to $J_1(\alpha a) = 0$, and a is the radius of the CDRA. The choice of cosΦ and sinΦ depends on the location of the feed. See [7] for a detailed explanation. The choice is referred from [7] and also explained there. Figure 1.2(a–o) shows the various modes and their field configurations; all are plots of the E-field orientation.

Theoretically, the resonant frequency calculation is simplified as [21]

$$f_0 = \frac{3\times 10^8}{2\pi\sqrt{\varepsilon_r}}\sqrt{2\left(\frac{\pi}{a}\right)^2 + \frac{\pi^2}{2h}} \tag{1.27}$$

As an example of a CDRA calculation [22], consider a dielectric material using Rogers TMM10I with a relative permittivity of $\varepsilon_r = 9.8$, radius $a = 2.5$ cm, and height, $h = 1.5$ cm, and placed on a ground plane of 10 cm × 10 cm. The antenna is excited by a coaxial probe of length 1 cm kept at the center. This CDRA resonates at 2.5 GHz, exciting the fundamental $TM_{01\delta}$ mode, which can be verified from equation (1.27). The CDRA offers a bandwidth of 14% (2.3–2.66 GHz) with a 3.8-dBi gain.

1.4 HDRA and Its Modes

The HDRA supports the body of revolution and hence the confined modes. The HDRA supports TE_{mnp} and TM_{mnp} modes, where m, n, and p represent the field variations along the radial (r), azimuth (Φ), and elevation (θ) directions, respectively. Since the dielectric air interface for an HDRA is simpler and has a smoother transition, a closed-form expression in terms of Green's function synthesis is possible. Hybrid modes have not been observed in HDRA so far. One of the reasons observed is that along the fully spherical or hemispherical body of revolution, observing longitudinal fields is not possible. This is also based on the right- and left-hand thumb rules, for elementary magnetostatics, where the E field and H field are always present in longitudinal directions, when all the field components along different axes are considered.

Since the confined modes condition, as stated in (1.1) and (1.2) is fulfilled, a simple current-source element to excite the HDRA can be considered for predicting the fields within. Unlike the RDRA, where to implement the DWM, a deliberate boundary condition needs to be imposed, the smoother transition of the fields make the Green's function synthesis possible for the modeling of the fields within HDRA.

The Green's function synthesis for an HDRA is

$$G_{TM/TE}^{Ar/Fr} = G_P + G_H \tag{1.28}$$

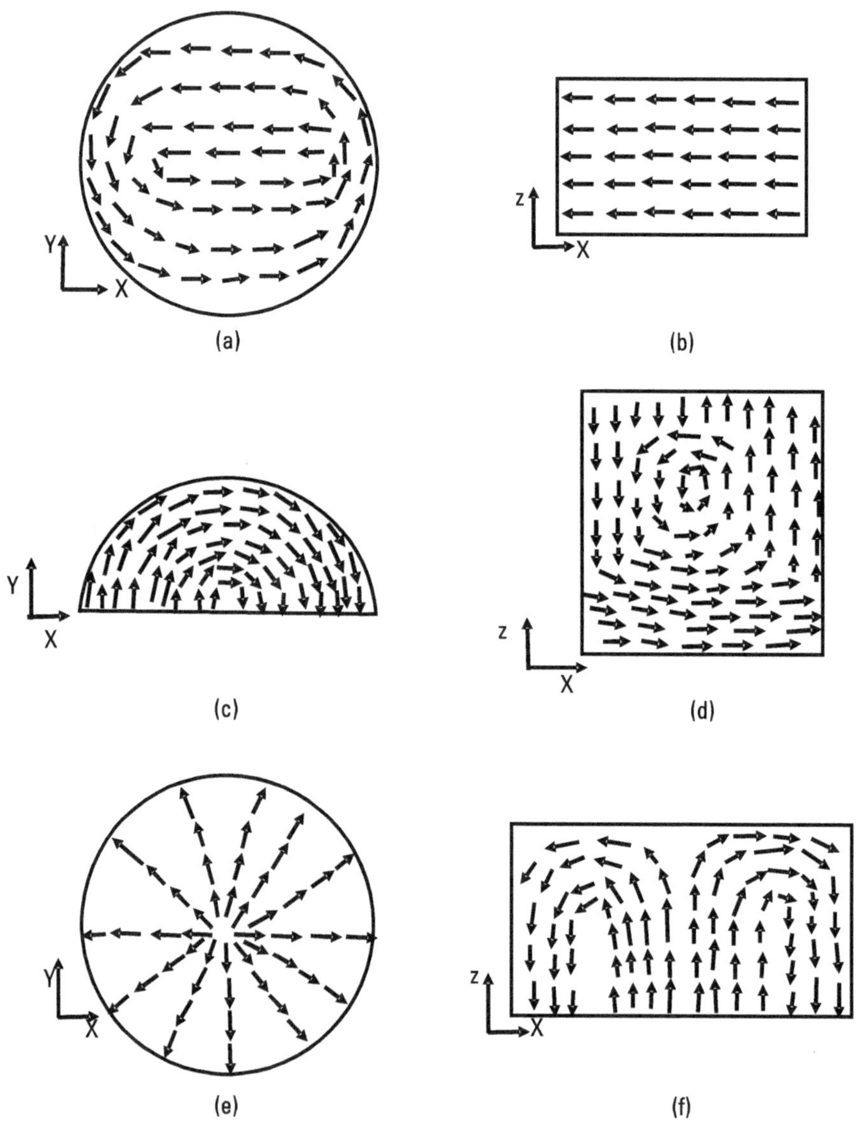

Figure 1.2 The modes of the CDRA: (a) $TE_{011+\delta}$ mode for the x-y plane, (b) $TE_{011+\delta}$ mode for the x-z plane, (c) $TE_{01\delta}$ mode for the x-y plane, (d) $TE_{01\delta}$ mode for the x-z plane, (e) $TM_{01\delta}$ mode for the x-y plane, (f) $TM_{01\delta}$ mode for the x-z plane, (g) $TM_{11\delta}$ mode for the x-y plane, (h) $TM_{11\delta}$ mode for the x-z plane, (i) $TM_{21\delta}$ mode for the x-y plane, (j) $TM_{21\delta}$ mode for the x-z plane, (k) $HE11\delta$ mode for the x-y plane, (l) $HE_{11\delta}$ mode for the x-z plane, (m) $HE_{12\delta}$ mode for the x-y plane, (n) $HE_{12\delta}$ mode for the x-z plane, and (o) $HE_{21\delta}$ mode for the x-z plane.

where Ar is the electric potential responsible for the generation of TM modes; F_r is the magnetic potential corresponding to the generation of TE modes; G_p is the particular solution of Green's function, which represents the source radiating in an unbounded dielectric medium; and G_H is the homogeneous solution

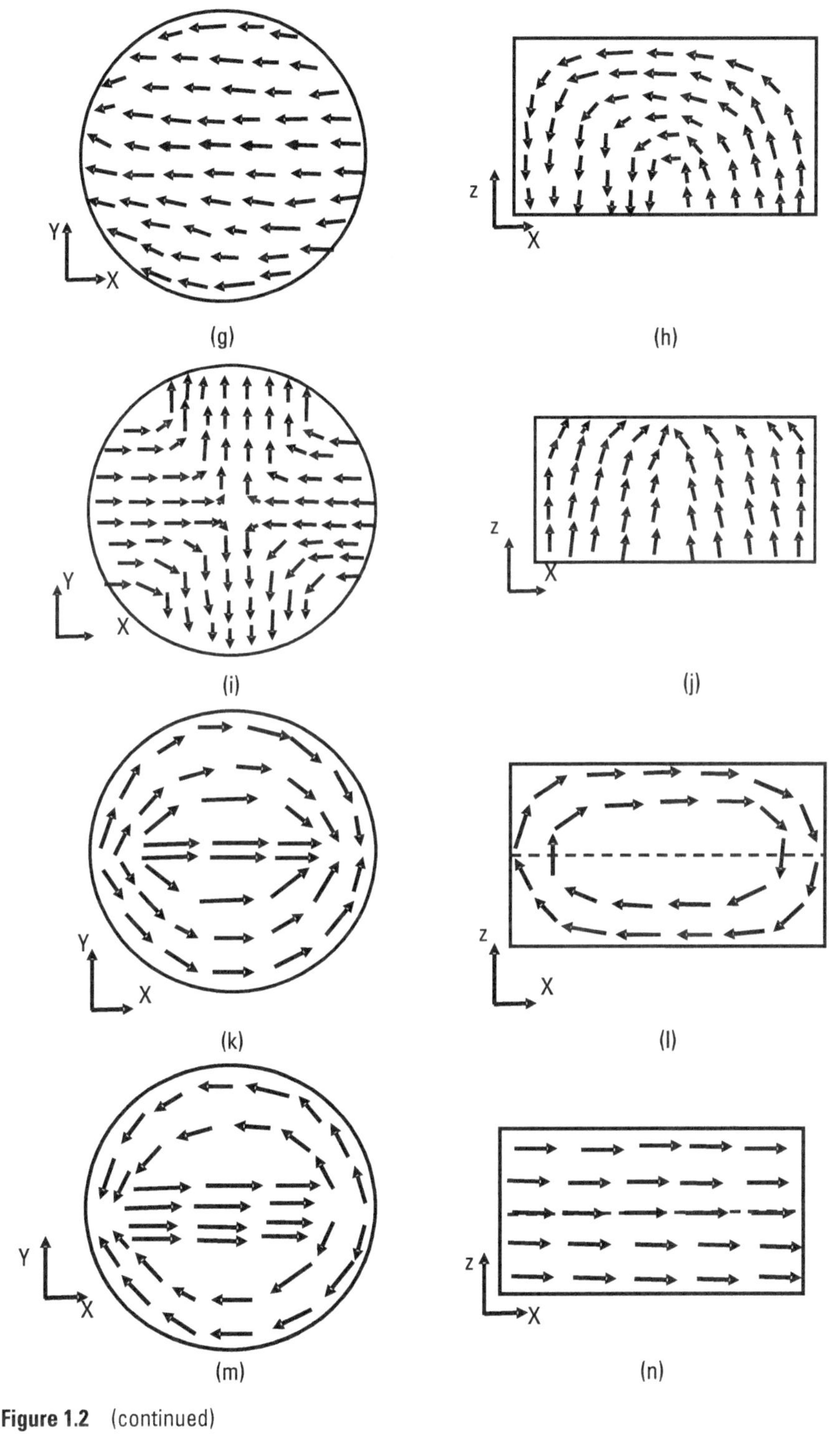

Figure 1.2 (continued)

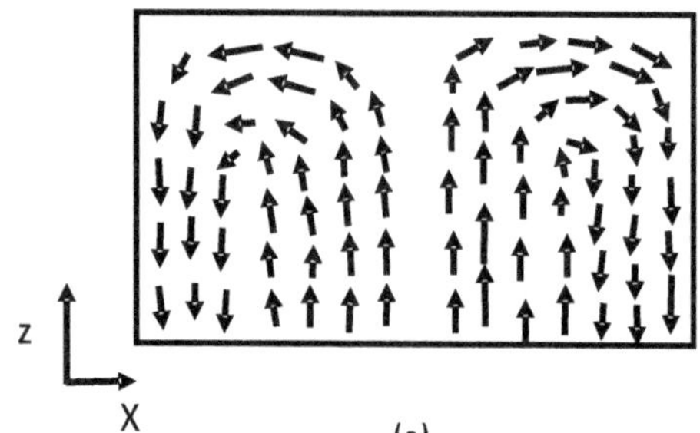

Figure 1.2 (continued)

of Green's function, which takes care of the dielectric boundary discontinuities at the dielectric/air interface. This is strictly in conjecture to the notations employed by [23–25]; it is also well deliberated and discussed in [7].

Each potential function is denoted by an infinite series of modal functions, and the modal coefficients are then obtained by matching boundary conditions at the source point and on the DRA surface, irrespective of the shape of the DRA [23–26]. This means that if an HDRA is excited by a probe feed that gives a $\hat{z}$-directed current, then the particular solution, G_P, involves solving the probe (conductor) and dielectric boundary condition where the dielectric is considered as an unbounded medium, and the homogeneous solution, G_H, is the solution to boundary condition over the dielectric-air interface. Image theory aids in calculating the equivalent currents, and thus Green's function is calculated [7].

If $\vec{r}(r,\theta,\varnothing)$ and $\vec{r'}(r,\theta,\varnothing)$ represent the field and source points respectively, then the equivalent Green's function for lowest-order TE mode (i.e., TE_{111} mode) is given by [7]

$$G_{TE_{111}} = \frac{-3k}{8\pi\omega\varepsilon rr'}\sin\theta\sin\theta'\cos(\varnothing-\varnothing') \left[\varnothing(kr')\Psi(kr)+\alpha_{TE}\widehat{J}_1(kr')\widehat{J}_1(kr)\right] \tag{1.29}$$

where

$$\phi(kr)=\begin{cases}\widehat{J}_1(kr'), r>r'\\ \widehat{H_1^{(2)}}, r<r'\end{cases} \tag{1.30}$$

$$\Psi(kr)=\begin{cases}\widehat{H_1^{(2)}}, r>r'\\ \widehat{J}_1(kr'), r<r'\end{cases} \tag{1.31}$$

$$\alpha_{TE} = \frac{-1}{\Delta_{TE}}\left[\widehat{H_1^{(2)}}(ka)\widehat{H_1^{(2)'}}(k_0a) - \sqrt{\varepsilon_r}\,\widehat{H_1^{(2)'}}(ka)\widehat{H_1^{(2)}}(k_0a)\right] \tag{1.32}$$

$$\Delta_{TE} = \widehat{J_1}(ka)\widehat{H_1^{(2)'}}(k_0a) - \sqrt{\varepsilon_r}\,\widehat{J_1}'(ka)\widehat{H_1^{(2)}}(k_0a) \tag{1.33}$$

In (1.29) to (1.33), $k_0 = \omega\sqrt{\mu_0\varepsilon_0}$, $k = \sqrt{\varepsilon_r}k_0$, $\widehat{J_1}(x)$ is the first-order spherical Bessel function of the first kind, and $\widehat{H_1^{(2)}}(x)$ is the spherical Hankel function of the second kind. The E field E_z due to the probe current J_z can be estimated by

$$E_z(\vec{r}) = G_{TE_{111}}\left(\vec{r},\vec{r'}\right)J_Z(z')dS' \tag{1.34}$$

$$J_Z(z') = J_0 \sin k\left(l - |z'|\right), -l \le z \le l \tag{1.35}$$

Equation (1.34) is calculated for the surface current flowing on the imaged probe surface of S_0; a is the radius of the hemisphere under consideration.

The input plots of the HDRA excited by a probe show that as the probe is displaced from the center of the circular base of the HDRA, the input impedance of the HDRA increases. This is because the probe current is r-directed at the center and so it excites TM modes, whereas when it is displaced, the probe excites the fundamental TE_{111} mode [23, 27].

For a generalized solution for the TM mode excited by a probe feed, the G_p and G_H in (1.28), are given as follows [7]

$$G_p(z,z') = \frac{-j}{\omega\varepsilon}\left(\frac{\partial^2}{\partial z^2} + k^2\right)\frac{e^{-jkR}}{4\pi R} \tag{1.36}$$

$$G_H(z,z') = \frac{-1}{4\pi\omega\varepsilon\, kz' z^2}\sum_{n=1}^{\infty} n(n+1)(2n+1)\alpha_n^{TM}\,\widehat{J_n}(kz')\widehat{J_n}(kz) \tag{1.37}$$

$$\alpha_n^{TM} = \frac{-\left[\widehat{H_n^{(2)'}}(ka)\widehat{H_n^{(2)}}(k_0a) - \sqrt{\varepsilon_r}\,\widehat{H_n^{(2)}}(ka)\widehat{H_1^{(2)'}}(k_0a)\right]}{\widehat{J_n'}(ka)\widehat{H_n^{(2)}}(k_0a) - \sqrt{\varepsilon_r}\,\widehat{J_n}(ka)\widehat{H_n^{(2)'}}(k_0a)} \tag{1.38}$$

where α_n^{TM} is the *TM*-mode reflection coefficient at the DRA boundary and $R = \sqrt{r_1^2 + (z - z')^2}$. The E field can be calculated by substituting the total Green's function as suggested in (1.34) for obtaining the fields for the *TM* modes.

The TE_{111} mode is the lowest order and hence the dominant mode of the HDRA. This mode radiates like a short magnetic dipole. The radiation pattern has a broadside nature, but the beamwidth is low. Conversely, the TM_{101} mode radiates like a short electric monopole antenna and has a null along the broadside direction of the radiation pattern.

The HDRA offers zero degrees of freedom, and so for a fixed frequency, the radius of the hemisphere is also fixed and cannot be altered. Figure 1.3 presents the various E field orientations for the TE and TM modes in an HDRA, with Figure 1.3(a) showing the TE_{111} mode [7] and Figure 1.3(b) showing the TM_{101} mode [7].

The resonant frequency of the HDRA is given by [27]

$$f_r = \frac{4.775 \times 10^7 \,\mathrm{Re}(ka)}{\sqrt{(\varepsilon_r)}r} \tag{1.39}$$

where f_r is the resonant frequency, ε_r is the dielectric constant of the HDRA, r is the radius of the hemisphere (in centimeters), and *Ka* is the wave number in the dielectric. For a dielectric material with $\varepsilon_r = 9.2$, the dissipation factor $\tan\delta = 0.0022$, and a density of 2.8 gm/cm^3, an HDRA with a radius of $r = 2.54$ cm (1") when excited at an offset probe feed of $x = 1.74$ cm from the center, resonates at 1.816 GHz with a bandwidth of 10.5%. The offset probe feed excites the fundamental TE_{111} mode of the HDRA [27].

1.5 Equilateral Triangular DRA and Its Modes

A detailed theoretical and experimental investigation of triangular DRAs has attracted the attention of DRA engineers. It is necessary to adopt the concepts

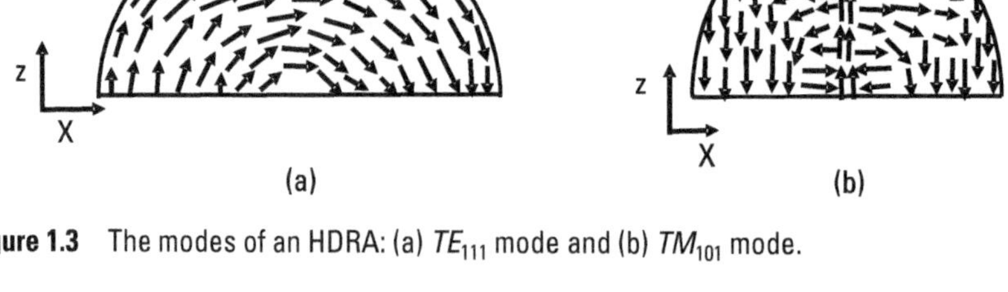

Figure 1.3 The modes of an HDRA: (a) TE_{111} mode and (b) TM_{101} mode.

of trilinear transformation and eigenvalues/eigenfunctions to predict the fields and resonant frequency of the equilateral triangular DRA (ETDRA). A significant contribution to this theoretical understanding can be found in [28–30].

For a conventionally placed TDRA on the ground plane, the top and bottom surfaces are observed with an imperfect magnetic conductor (IPMC), and the side wall, which is rectangular in nature, is modeled with a perfect magnetic conductor (PMC) boundary. Trilinear transformation is used as defined in [30] to generate a local coordinate system inside the equilateral triangular section. The equilateral triangular section is symmetric with respect to the *x*-axis as shown in Figure 1.4.

However, after applying trilinear transformation, the symmetry gets shifted to the *y*-axis. The approximate solution of the eigenfunction for even modes TM^z mode is

$$\psi = F(x,y) \times H(z) \tag{1.40}$$

where

$$\begin{aligned} F(x,y) &= \cos\frac{2\pi l}{\sqrt{3}a}(x+R)\cos\left(\frac{2\pi(m-n)}{3a}y\right) \\ &+ \cos\frac{2\pi m}{\sqrt{3}a}(x+R)\cos\left(\frac{2\pi(n-l)}{3a}y\right) \\ &+ \cos\frac{2\pi n}{\sqrt{3}a}(x+R)\cos\left(\frac{2\pi(l-m)}{3a}y\right) \end{aligned} \tag{1.41}$$

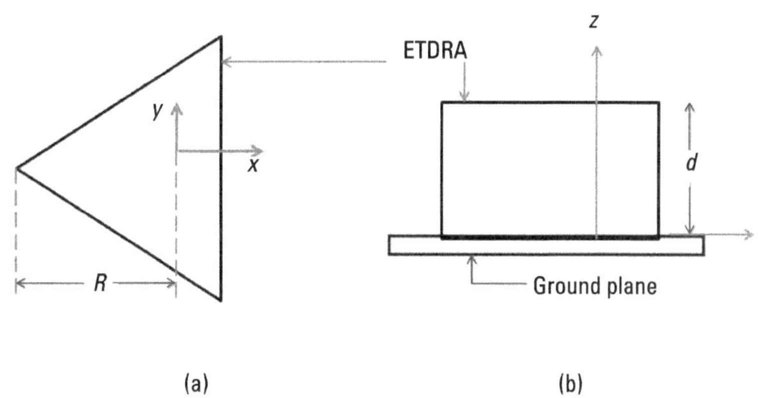

Figure 1.4 The ETDRA geometry: (a) top view and (b) side view.

and $H(z)$ is expected to be a standing wave and not a traveling wave for modeling ETDRA as a resonator. m, n, l, and p are integers such that $l = -(m+n), R = \frac{a}{\sqrt{3}}, a$ is the side of the ETDRA..

For odd modes, the $F(x,y)$ is given as

$$F(x,y) = \cos\frac{2\pi l}{\sqrt{3}a}(x+R)\sin\left(\frac{2\pi(m-n)}{3a}y\right) + \cos\frac{2\pi m}{\sqrt{3}a}(x+R)\sin\left(\frac{2\pi(n-l)}{3a}y\right) + \cos\frac{2\pi}{\sqrt{3}a}(x+R)\sin\left(\frac{2\pi(l-m)}{3a}y\right) \tag{1.42}$$

However, for a TDRA placed on the ground plane, odd modes are not observed. Equation (1.42) is valid for source-free and isolated ETDRA only. In order to evaluate the standing field $H(z)$ in the ETDRA, two cases for the boundary conditions are considered separately. First, it is assumed that the source-free isolated ETDRA is bounded by PMC along its height also. This model is termed the PMC model. Also, it is assumed that the maximum fields are confined within the source-free isolated ETDRA and that the fields outside the isolated ETDRA are decaying exponentially. This model is termed the IPMC model.

Since the odd mode is not possible, the final standing-wave expression is

$$H(z) = \cos(k_z z) \tag{1.43}$$

where k_z is the solution of

$$k_z \tan\left(\frac{k_z h}{2}\right) - \varepsilon_r \gamma_z = 0 \tag{1.44}$$

Thus, the eigenfunction for the isolated ETDRA is

$$\psi^{e}_{mnp} = \begin{bmatrix} \cos\frac{2\pi l}{\sqrt{3}a}(x+R)\cos\left(\frac{2\pi(m-n)}{3a}y\right) \\ +\cos\frac{2\pi m}{\sqrt{3}a}(x+R)\cos\left(\frac{2\pi(m-l)}{3a}y\right) \\ +\cos\frac{2\pi}{\sqrt{3}a}(x+R)\cos\left(\frac{2\pi(l-m)}{3a}y\right) \end{bmatrix} \times \begin{cases} \sin\left[k_z(d-z)\right] \textit{for PMC} \\ \cos\left[k_z(d+\beta)\right] \textit{for IPMC} \end{cases} \quad (1.45)$$

The fundamental mode is the TM_{101} mode for the ETDRA. For any arbitrary given T^{z}_{mnp} mode excited in the ETDRA, the resonant frequency is

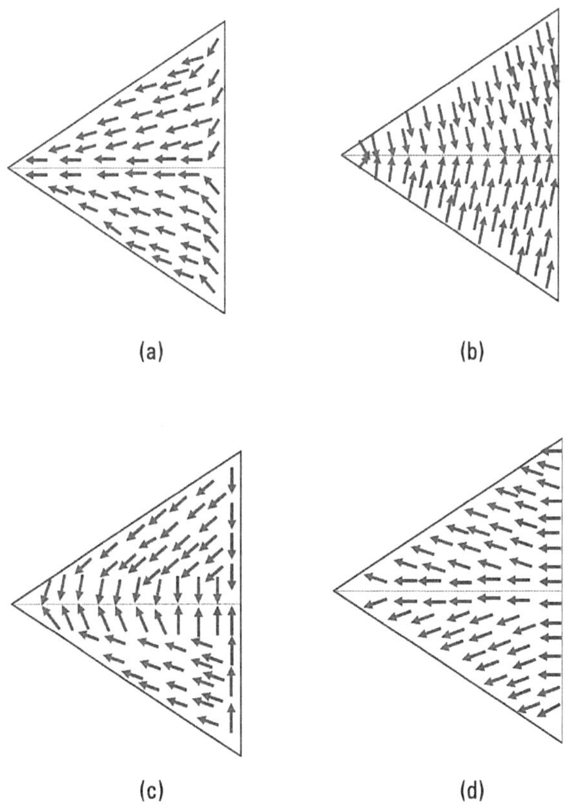

Figure 1.5 The TM_{101} mode of ETDRA: (a) even-mode E field, (b) even-mode H field, (c) odd-mode E field, and (d) odd-mode H field.

$$f_r = \frac{c}{2\pi\sqrt{\varepsilon_r}} \times \left[\left(\frac{4\pi}{3a} \right)^2 \left(m^2 + mn + n^2 \right) + k_z^2 \right]^{1/2} \tag{1.46}$$

where the symbols have their usual meanings as discussed. Based on the equivalence principle the far-field pattern corresponding to the TM_{101} mode can also be estimated as in [30].

The plot of the TM_{101} mode of the ETDRA is as shown in Figure 1.5. This is as referred to in [30].

1.6 Conclusion

This chapter has provided a basic understanding of the modes and the resonant frequency calculation of various DRA configurations. In addition, it clearly explained the mode nomenclature and the anatomy of modes, with figures and examples. Subsequent chapters present various investigations into DRAs for research and commercial applications.

References

[1] Chatterjee, R., *Dielectric and Dielectric Loaded Antennas,* Wiley, 1985.

[2] Long, S. A., M. W. McAllister, and L. C. Shen, "Rectangular Dielectric Resonator Antenna," *IEEE Electronics Letters,* Vol. 19, March 1983, pp. 218–219.

[3] Long, A., M. W. McAllister, and L. C. Shen, "The Resonant Cylindrical Dielectric Cavity Antenna," *IEEE Trans Antennas Propag.*, Vol. 31, No. 3, May 1983, pp. 406–412.

[4] McAllister, M. W., and S. A. Long, "Resonant Hemispherical Dielectric Antenna," *Electron. Lett.*, Vol. 20, No. 16, 1984, pp. 657–659.

[5] Mongia, R. K., and A. Ittipiboon, "Theoretical and Experimental Investigations on Rectangular Dielectric Resonator Antennas," *IEEE Transactions on Antennas and Propagation,* Vol. 45, No. 9, 1997, pp. 1348–1356.

[6] Van Bladel, J., "On the Resonances of a Dielectric Resonator of Very High Permittivity," *IEEE Transactions on Microwave Theory and Techniques,* Vol. MTT-23, Feb. 1975, pp. 199–208.

[7] Mukherjee, B., P. Patel, and J. Mukherjee, "A Review of the Recent Advances in Dielectric Resonator Antennas," *Journal of Electromagnetic Waves and Applications, Taylor & Francis,* Vol. 34, Issue 9, 2020, pp. 1095–1198.

[8] Okaya, A. K., and L. F. Barash, "The Dielectric Microwave Resonator," *Proceedings of IRE,* Vol. 50, 1962, pp. 2081–2092.

[9] Zou, M., and J. Pan, "Investigation of Resonant Modes in Wideband Hybrid Omnidirectional Rectangular Dielectric Resonator Antenna," *IEEE Transactions on Antennas and Propagation,* Vol. 63, No. 7, 2015, pp. 3272–3275.

[10] Kiran, V., S. Dileep, and B. Mukherjee, "Compact Embedded Dual-element Rectangular Dielectric Resonator Antenna Combining Sierpinski and Minkowski Fractals," *IEEE Transactions on Components, Packaging and Manufacturing Technology,* Vol. 7, No. 5, 2017, pp. 786–791.

[11] Gupta, J., B. Mukherjee, and N. Gupta, "A Novel Tetraskelion Dielectric Resonator Antenna for Wideband Applications," *Microwave and Optical Technology Letters,* Wiley, Vol. 57, No. 12, 2015, pp. 2781–2786.

[12] Patel, P., B. Mukherjee, and J. Mukherjee, "Wideband Circularly Polarized Rectangular Dielectric Resonator Antenna Using Square Shaped Slots," *IEEE Antenna and Wireless Propagation Letters,* Vol. 15, 2016, pp. 1309–1312.

[13] Patel, P., B. Mukherjee, and J. Mukherjee, "A Compact Wideband Rectangular Dielectric Resonator Antenna Using Perforations and Edge Grounding," *IEEE Antenna and Wireless Propagation Letters,* Vol. 14, 2015, pp. 490–493.

[14] Luk, K., and M. K. W. Leung, *Dielectric Resonator Antennas,* England: Research Studies Press Ltd., 2003.

[15] Petosa, A., *Dielectric Resonator Antenna Handbook,* Norwood, MA: Artech House, 2007.

[16] Petosa, A., and A. Ittipiboon, "Dielectric Resonator Antennas: A Historical Review and the Current State of the Art," *IEEE Antenna and Propagation Magazine,* Vol. 52, No. 5, 2010, pp. 91–116.

[17] Sethares, J. C., and J. Nauman, "Design of Microwave Dielectric Resonators," *IEEE Transactions on Microwave Theory and Techniques,* Vol. 14, No. 1, 1966, pp. 2–7.

[18] Verplanken, M., and J. Van Bladel, "The Magnetic-Dipole Resonance of Ring Resonators of Very High Permittivity," *IEEE Transactions on Microwave Theory and Techniques,* Vol. 27, No. 4, 1979, pp. 328–332.

[19] Kafez, D., and P. Guillon (eds.), *Dielectric Resonators,* Norwood, MA: Artech House, 1986.

[20] N. Gupta, S. K. Rout, and K. Sivaji, "Characteristics of Cylindrical Dielectric Resonator Antenna," *Microwave Review,* Vol. 15, No. 2, 2009, pp. 29–32.

[21] Dileep S., Venkat Kiran, B. Mukherjee, "Compact Bi-cone Dielectric Resonator Antenna for Ultra-Wideband Applications—A Novel Geometry Explored," *Electromagnetics,* Taylor & Francis, Vol. 37, Issue 7, 2017, pp. 471–481.

[22] Luk, K. M., and K. W. Leung (eds.), *Dielectric Resonator Antennas,* Baldock, Hertforshire, England: Research Studies Press Ltd., 2003.

[23] Kakade, A. B., and B. Ghosh, "Efficient Techniques for the Analysis of Microstrip Slot-Coupled Hemispherical Dielectric Resonator Antenna," *IEEE Antennas and Wireless Propagation Letters,* Vol. 7, 2008, pp. 332–336.

[24] Kakade, A. B., and B. Ghosh, "Mode Excitation in the Coaxial Probe Coupled Three-Layer Hemispherical Dielectric Resonator Antenna," *IEEE Transactions on Antennas and Propagation,* Vol. 59, 2011, pp. 4463–4469.

[25] Bhattacharya, et al., "Evaluation of Efficient Green's Functions for Spherically Stratified Media," *IEEE Transactions on Antennas and Propagation,* Vol. 66, 2018, pp. 1613–1618.

[26] Mukherjee, B., P. Patel, and J. Mukherjee, "A Novel Cup-shaped Inverted Hemispherical Dielectric Resonator Antenna for Wideband Applications," *IEEE Antenna and Wireless Propagation Letters,* Vol. 12, 2013, pp. 1240–1243.

[27] Kishk, A. A., "A Triangular Dielectric Resonator Antenna Excited by a Coaxial Probe," *Microwave and Optical Technology Letters,* Wiley, Vol. 30, No. 5, 2001, pp. 340–341.

[28] Maity, S., and B. Gupta, "Comments on a 'A Triangular Dielectric Resonator Antenna Excited by a Coaxial Probe'," *Microwave and Optical Technology Letters,* Wiley, Vol. 54, No. 6, 2012, pp. 1548.

[29] Maity, S., and B. Gupta, "Theoretical Investigations on Equilateral Triangular Dielectric Resonator Antenna," *IET Microwave and Antenna Propagation,* Vol. 11, Issue 2, 2017, pp. 184–192.

[30] Schelkunoff, S. A., *Electromagnetic Waves,* D. Van Nostrand Company, Inc., 1957.

2

DRA Bandwidth Enhancement Techniques

The demand for numerous wireless applications in recent years has spawned the need for an efficient, wideband system to cover the required bandwidth; a wideband antenna fulfills this requirement.

Wideband antennas are useful for many applications, including satellite, radar, mobile communication, and weather monitoring. Current portable and fixed devices require a high–data rate switching transition from one frequency range to another frequency range without any delay. Wideband antennas meet this demand without any performance degradation. Various configurations of the DRA can cover broadband, wideband, dual-band, multiband, and ultra-wideband frequency ranges.

This chapter explores various design techniques to achieve the wideband frequency range, classifying the techniques into various subsections, such as feeding excitation methods, stacking various permittivity layers, metallic-loading with DRAs, compact and low-profile geometry, and fractal and reshaped CDRA geometry.

2.1 Various Feeding Methods

Several feeding techniques have been proposed to excite the DRA to achieve a wide-impedance bandwidth. For a single DRA, wideband, multiband, and ultra-wideband frequency can be achieved with proper excitation. Energy coupling between feed networks to a DRA plays an important role in achieving a wideband frequency response, as do many other feeding methods, including coaxial probe, microstrip line, slot aperture, and conformal feed. In addition, maximum energy is coupled from the feeding network to the DRA when the coupling factor is approximately 1. The Q factor is dependent on the coupling factor. If maximum energy is coupled, then the overall Q factor of a loaded feed network decreases, and the bandwidth is enhanced simultaneously [1] as shown by

$$Q_l = \frac{Q}{1+x} \tag{2.1}$$

where Q_l is the loaded Q factor of the DRA, and x is the coupling factor, which depends on the coupling mechanism. The loaded Q factor plays an important role in enhancing the bandwidth. Various feeding networks are studied in this section to obtain wideband frequency response.

2.1.1 Coaxial Probe Feed

Figure 2.1 shows coaxial probe coupling for RDRA, CDRA, and HDRA. The central conductor of the coaxial probe is extended (probe) inside the DRA for efficient coupling and perfect impedance matching. Figure 2.2 illustrates E and H field source generation in DRAs. Feed location and feed length are prominent elements of this technique, because both the coupling factor and impedance match depend on them.

Similarly, when the DRA is suitably excited, it generates suitable electromagnetic (EM) modes leading to a different far-field radiation pattern. These modes can be merged to provide a wide bandwidth of operation. However, proper selection of feed position through holes drilled from the ground plane and insertion of metal of coaxial pin, respectively, in the DRA is quite difficult. The drilling of holes and insertion of a metal pin in DRAs leads to fabrication complexity.

2.1.1.1 Designing an RDRA Through Coaxial Feed

A wideband RDRA is achieved by proper selection of design parameters such as height (h), width (w), length (l), dielectric material (ε_r), feed length (l_p), and position. TE and TM modes are investigated in rectangular DRA [2, 3] after applying the perfect magnetic wall conditions. If the rectangular DRA is placed on the ground plane, typically only TE modes are excited, and its radiation resembles a short magnetic dipole. For the TE mode, the Q factor can be obtained by calculating the total stored energy (W_e) and power radiated (P_{rad}) through the DRA boundary [4] as

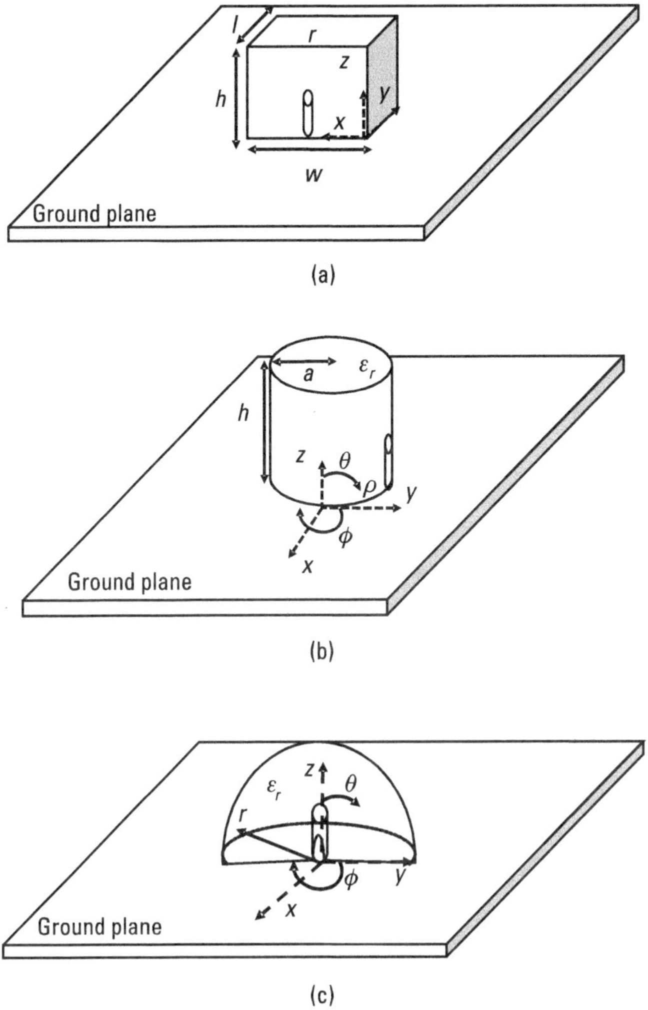

Figure 2.1 RDRA, CDRA, and HDRA excited by a probe-feeding network.

$$Q = \frac{2\omega W_e}{P_{rad}} \tag{2.2}$$

Rectangular DRA resonant frequency is calculated by the curve-fitting method and normalized *F* frequency can be calculated as

$$f = \frac{15F}{w * \pi * \sqrt{\varepsilon_r}} \tag{2.3}$$

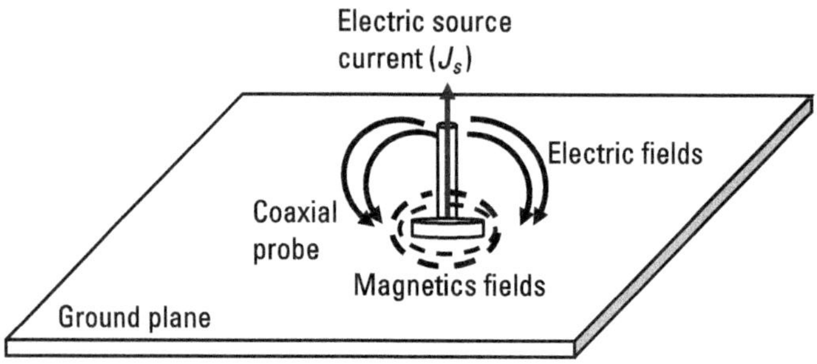

Figure 2.2 E and H fields source generation.

$$F = a_0 + a_1\left(\frac{w}{2h}\right) + a_2\left(\frac{w}{2h}\right)^2 \tag{2.3a}$$

$$a_0 = 2.57 - 0.8\left(\frac{l}{2h}\right) + 0.42\left(\frac{l}{2h}\right)^2 - 0.05\left(\frac{l}{2h}\right)^3 \tag{2.3b}$$

$$a_1 = 2.71\left(\frac{l}{2h}\right)^{-0.282} \tag{2.3c}$$

$$a_{2=}0.16 \tag{2.3d}$$

The $TE^z_{11\delta}$ dominant mode can be excited when the coaxial probe is placed adjacent to the rectangular DRA. Figure 2.3 illustrates field excitation through probe feeding.

2.1.1.2 Designing a CDRA Through Coaxial Feed

A wideband CDRA can be designed based on the parameters of radius (a), height (h), dielectric material (ε_r), feed length, and position. For calculating the Q factor of CDRA, k_0a (wave number) can be calculated by using a rigorous numerical method with proper selection of height (h)-to-radius (a) ratio of dielectric material. The resonant frequency of the desired band depends on k_0a and aspect ratio (a/h). To excite various modes through the coaxial feed method of CDRA, the Q factor, k_0a, and an extensive resonant frequency design equation are taken from [3, 5].

For the $HE_{11\delta}$ mode, the following is used to calculate the Q and k_0a.

$$Q = 0.01007(\varepsilon_r)^{1.30}\left(a/h\right)\left\{1 + 100e^{-2.05\left[0.5\left(a/h\right) - 0.0125\left(a/h\right)^2\right]}\right\} \tag{2.4}$$

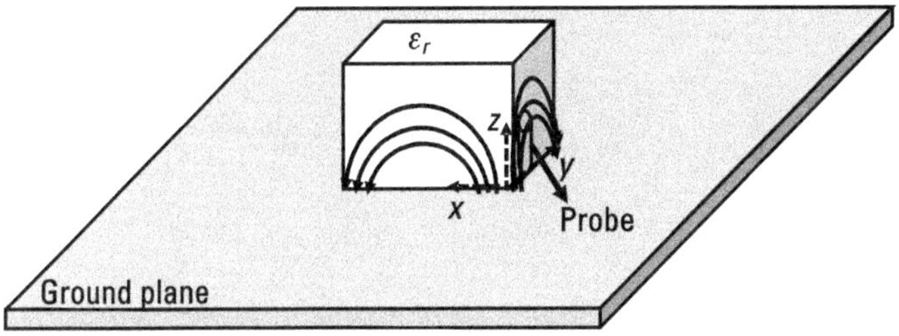

Figure 2.3 $TE^{z}_{11\delta}$ mode field distribution in an RDRA excited by a probe-feeding network.

$$k_0 a = \frac{6.324}{\sqrt{\varepsilon_r + 2}}\left\{0.027 + 0.36\left(\frac{a}{2h}\right) + 0.02\left(\frac{a}{h}\right)^2\right\} \tag{2.5}$$

The radiating $HE_{11\delta}$ mode can be excited when a coaxial probe is placed adjacent to the CDRA. Figure 2.4 shows a probe feeding a CDRA for excitation of $HE_{11\delta}$ mode.

For the $TE_{01\delta}$ and $TE_{011+\delta}$ mode, the following is used to calculate the Q factor and k_0a for the $TE_{01\delta}$ mode.

$$Q = 0.078192(\varepsilon_r)^{1.27}\left\{\begin{matrix} 1 + 17.31\left(\frac{h}{a}\right) - 21.57\left(\frac{h}{a}\right)^2 \\ +10.86\left(\frac{h}{a}\right)^3 - 1.98\left(\frac{h}{a}\right)^4 \end{matrix}\right\} \tag{2.6}$$

$$k_0 a = \frac{2.327}{\sqrt{\varepsilon_r + 1}}\left\{1 + 0.2123\left(\frac{a}{h}\right) - 0.00898\left(\frac{a}{h}\right)^2\right\} \tag{2.7}$$

Similarly, for the $TE_{011\delta}$ mode Q factor and k_0a are calculated as follows.

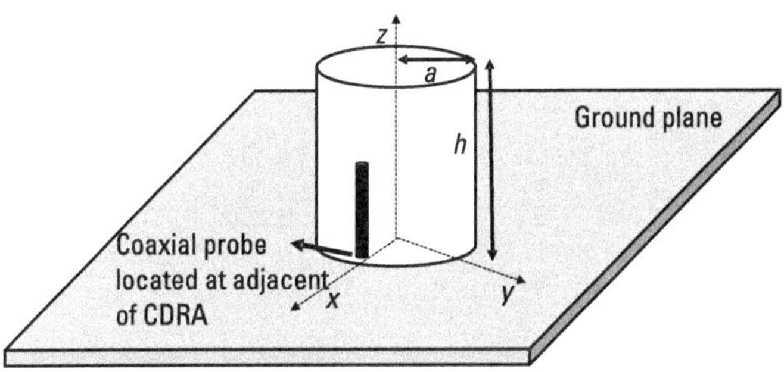

Figure 2.4 $HE_{11\delta}$ mode in a CDRA excited by a probe-feeding network [3].

$$Q = 0.3628(\varepsilon_r)^{2.38}\begin{Bmatrix} -1 + 7.81\left(\frac{h}{a}\right) - 5.858\left(\frac{h}{a}\right)^2 \\ +1.277\left(\frac{h}{a}\right)^3 \end{Bmatrix} \tag{2.8}$$

$$k_0 a = \frac{2.208}{\sqrt{\varepsilon_r + 1}}\left\{1 + 0.7013\left(\frac{a}{h}\right) - 0.002713\left(\frac{a}{h}\right)^2\right\} \tag{2.9}$$

The radiating $TE_{01\delta}$ and $TE_{011+\delta}$ mode can be excited by the half-cylindrical DRA, when a coaxial probe is fixed at the adjacent edge of the half CDRA. Figure 2.5 shows a probe feeding a half CDRA for exciting the $TE_{01\delta}$ mode.

For the $TM_{01\delta}$ mode, a magnetic wall boundary and computational method is used to calculate the accurate value of the wave number (i.e., k_0a). This leads to the following for k_0a

$$k_0 a = \frac{\sqrt{(3.83)^2 + \left(\frac{\pi a}{2h}\right)}}{\sqrt{\varepsilon_r + 2}} \tag{2.10}$$

The $TM_{01\delta}$ mode can be excited in CDRA, when a coaxial probe is located at the center of the CDRA. Figure 2.6 shows a probe feeding a CDRA for excitation of the $TM_{01\delta}$ mode.

CDRA resonant frequency is calculated by a rigorous numerical method, and the resonant frequency (f_r) formula is

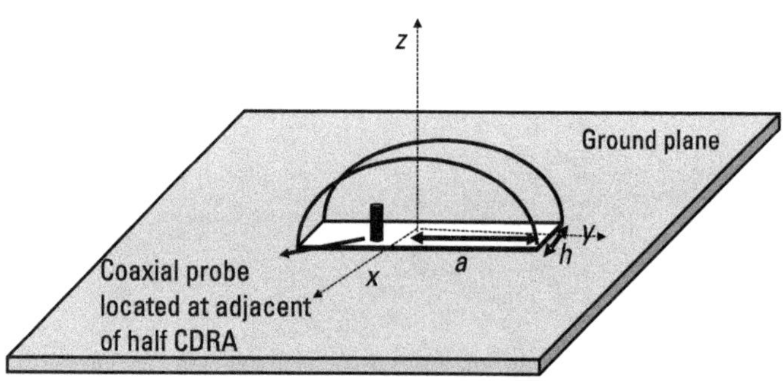

Figure 2.5 $TE_{01\delta}$ mode in a CDRA excited by a probe-feeding network [3].

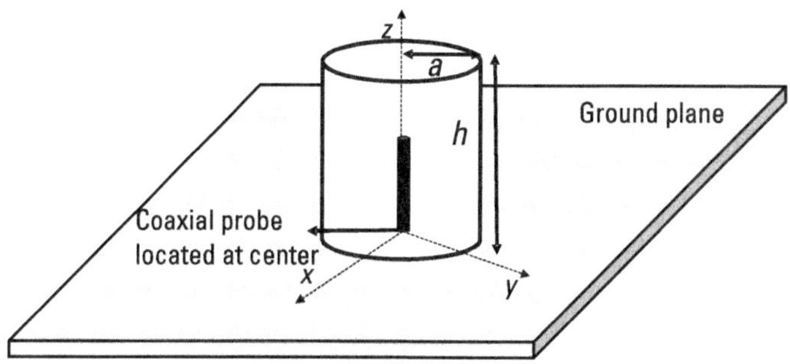

Figure 2.6 $TM_{01\delta}$ mode in a CDRA excited by a probe-feeding network.

$$f_r = \frac{(k_0 a) * c}{2\pi a} \tag{2.11}$$

where c is the light velocity, $k_0 a$ is the wave number, and a is the cylinder radius.

2.1.1.3 Designing an HDRA Through Coaxial Feed

An HDRA is constructed by the proper selection of the dielectric material (ε) and the design parameter of radius (r). Figure 2.7 depicts the HDRA geometry. Hemispherical DRA is placed on a copper ground plane and excited through a coaxial probe. If the HDRA is fixed on the conducting surface, the height of the HDRA is doubled because of the application of the image theory concept. TE modes are generated when the radial component of the E field is zero ($E_r = 0$); similarly TM modes are generated when the radial component of the H field is zero ($H_r = 0$). In HDRA, the field varies in the directions of the radial (r), azimuth (φ), and elevation (θ). For the fundamental TE_{111} and TM_{101} mode, the Q factor, $k_0 a$, and resonant frequency are calculated through (2.12) to (2.15) [6–7].

For the TE_{111} mode, the characteristic equation (2.12) to determine the Q factor and resonant frequency is

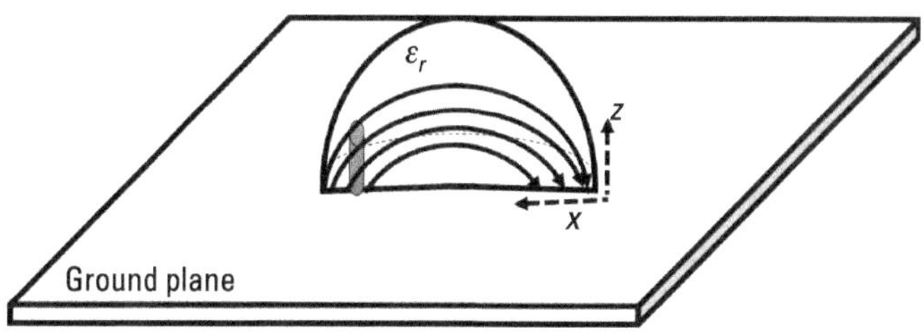

Figure 2.7 The TE_{111} mode and field distribution in an HDRA excited by a probe-feeding network.

$$\frac{J_{\frac{1}{2}}\left(k_0 a\sqrt{\varepsilon_r}\right)}{J_{\frac{3}{2}}\left(k_0 a\sqrt{\varepsilon_r}\right)} = \frac{H_{1/2}^{(2)}\left(k_0 a\right)}{\sqrt{\varepsilon_r} H_{3/2}^{(2)}\left(k_0 a\right)} \tag{2.12}$$

In (2.12), J_n is first type of the spherical Bessel function of the *n*th order, $H_n^{(2)}$ is the spherical second-order Hankel function of the *n*th order, and ε_r is the dielectric material permittivity. To obtain the resonant frequency, (2.12) is solved to yield

$$f_{GHz} = \frac{4.7713 Re\left(k_0 a\right)}{r} \tag{2.13}$$

where *f* is the resonant frequency, $k_0 a$ is the wave number, and r is the radius of the hemisphere. The Q factor is calculated from

$$Q = \frac{Re\left(k_0 a\right)}{2 Im\left(k_0 a\right)} \tag{2.14}$$

$Re(k_0 a)$ and the Q factor are calculated as follows.

$$Re\left(k_0 a\right) = 2.8316\varepsilon_r^{-0.47829} \tag{2.15a}$$

$$Q = 0.08 + 0.796\varepsilon_r + 0.01226\varepsilon_r^2 - 0.00003\varepsilon_r^3 \tag{2.15b}$$

The TE_{111} mode can be excited in an HDRA, when a coaxial probe is located at an offset from the center of the HDRA. Figure 2.7 shows a probe feeding an HDRA for excitation of TE_{111} mode.

The TM_{101} mode for TM_{101}, resonant frequency, and the Q factor is calculated from the transcendental equation

$$\frac{1}{k_0 a\sqrt{\varepsilon_r}} - \frac{J_{\frac{1}{2}}\left(k_0 a\sqrt{\varepsilon_r}\right)}{J_{\frac{3}{2}}\left(k_0 a\sqrt{\varepsilon_r}\right)} = \frac{\sqrt{\varepsilon_r}}{k_0 a} - \sqrt{\varepsilon_r} \bullet \frac{H_{1/2}^{(2)}\left(k_0 a\right)}{H_{3/2}^{(2)}\left(k_0 a\right)} \tag{2.16}$$

For high dielectric material permittivity ($\varepsilon_r > 20$), the Q factor can be calculated from

$$Q = 2.621 - 0.574\varepsilon_r + 0.02812\varepsilon_r^2 + 0.0000259\varepsilon_r^3 \tag{2.16a}$$

Similarly, for low dielectric material permittivity ($\varepsilon_r < 20$), the Q factor can be calculated from

$$Q = 0.723 - 0.932\varepsilon_r + 0.0956\varepsilon_r^2 + 0.00403\varepsilon_r^3 - 0.000005\varepsilon_r^4 \quad (2.16b)$$

and

$$Re(k_0 a) = 4.47226\varepsilon_r^{-0.505} \quad (2.16c)$$

The TM_{101} mode can be excited in an HDRA, when the coaxial probe is located at the center of the HDRA. Figure 2.8 shows a probe feeding an HDRA for excitation of the TM_{101} mode.

Researchers have explored various types of DRAs for wideband applications that are excited by coaxial probe feed networks. In [8] an RDRA with a concave ground plane is investigated for achieving a frequency band of 3.28–5.77 GHz. A concave ground plane has air introduced between the DRA and the ground plane, which improves the bandwidth performance. $TE_{11\delta}$ mode is generated when the DRA is excited through a coaxial probe that is located at an offset from the center. Similarly in [9], for achieving wide bandwidth, a coaxial feed with bevel patch feeding is used to excite the RDRA. Due to the coaxial bevel feeding technique, an air gap in the DRA leading to the feeding network reduces the effective dielectric constant and the Q factor of the antenna. Figure 2.9 depicts the DRA with a feeding network configuration. It is constructed using Rogers TMM 10i dielectric material ($\varepsilon_r = 9.8$), and it covers a wide frequency range from 2.6 GHz to 11 GHz for numerous wireless applications.

In [10], wideband is achieved by a single probe feed excitation to a DRA; note that the author examined the radius-to-height (*r*/*h*) ratio and the length-to-height (*l*/*h*) ratio, respectively, for the cylindrical and rectangular geometries to obtain the wideband frequency response. The proper selection of the radius-to-height ratio of the CDRA offers a dual mode of frequency of operation. $HE_{11\delta}$ and $HE_{11\Delta}$ (also known as TM_{110} and TM_{111}, respectively) [10] modes are generated, respectively, corresponding to the resonant frequency. Its dual

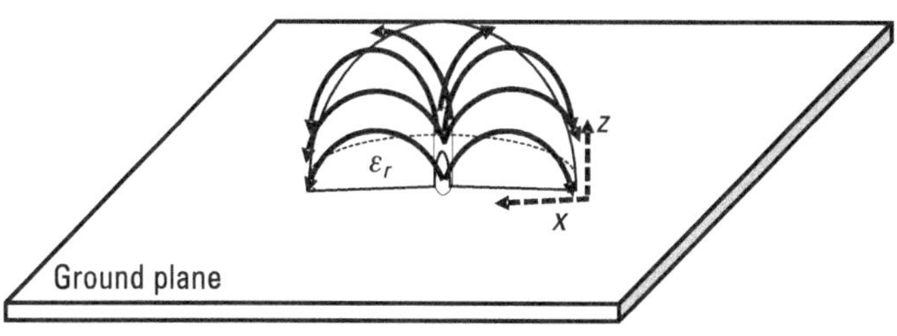

Figure 2.8 The TM_{101} mode in a field distribution in an HDRA excited by a probe-feeding network.

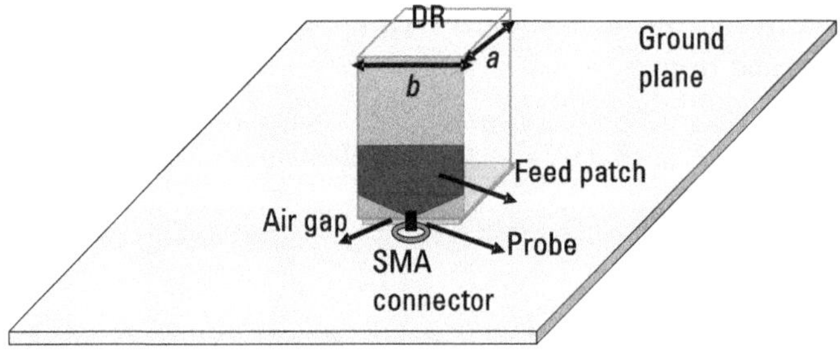

Figure 2.9 An RDRA with a feeding network configuration.

modes are merged and offer a wideband frequency response. This cylindrical DRA is constructed by Eccostock HiK500 dielectric material $\varepsilon_r = 10$, 12, 14, 16, and 18 with a cylinder radius of 12.7 mm, and the prototype is experimentally verified. This structure obtained the best results at ($a/h = 0.41$) and $\varepsilon_r = 10$. Figure 2.10 shows the single probe feed–excited CDRA geometry and a plot of percentage bandwidth versus a/h for various $\varepsilon_r = 10$–18.

In [11], a coaxial feed with a conformal strip-fed CDRA is investigated for bandwidth enhancement. However, this antenna covers a 1.6–3.2-GHz (66%) impedance bandwidth and accommodates UMTS, WLAN, DCS, and PCS wireless bands, respectively. Air is introduced within the DRA, which reduces the effective dielectric constant of the material and enhances the bandwidth of antenna. This antenna is fabricated using Eccostock ceramic material ($\varepsilon_r = 10$), and a metallic strip is connected to the center part of the coaxial pin of the SMA

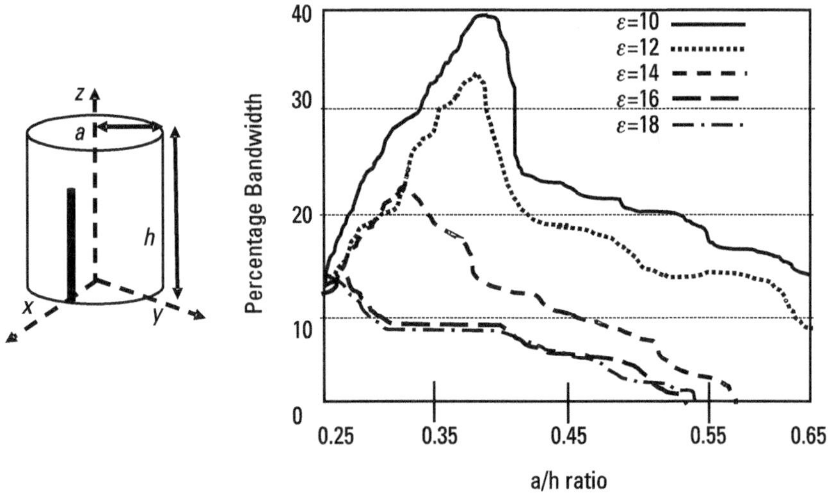

Figure 2.10 (a) CDRA geometry excited by a probe feed placed on the x-axis and (b) CDRA bandwidth versus a/h ratio plot for varying $\varepsilon_r = 10^{-18}$ [10].

connector. Figure 2.11 shows the top and side views of the antenna geometry. In [12], laterally placed cylindrical DRA triangular notches are presented for enhancement of the bandwidth. In order to design the antenna, notches are created at the edges, and the CDRA is placed laterally. This antenna is excited through a single coaxial feed that offers a 4.5–10.1-GHz (76.7%) wide impedance bandwidth. Laterally placed CDRA on the ground plane reduced the base area contact of the CDRA to the ground plane. As a result, more current gets coupled through CDRA, thereby enhancing the radiation losses and contributing to the enhancement of the bandwidth. To further improve the antenna bandwidth, two triangular notches are created at the edges of a laterally placed CDRA. By introducing the notches, the effective permittivity of the CDRA is decreased, thus improving the overall bandwidth performance. $HE_{11\delta}$-, $HE_{12\delta}$-, and $HE_{21\delta}$-like modes are generated corresponding to the resonant frequencies of 5.12 GHz, 7.28 GHz, and 9.48 GHz, respectively. Figure 2.12 shows the fabricated structure.

Similarly, in [13], a wideband two-element, half-split, cylindrical DRA excited through a single probe element is investigated. This antenna offers a 66% frequency band for WLAN and WiMAX applications. In [14] for achieving a wide bandwidth and a monopole-like radiation pattern, four-element CDRA arrangements are studied. A small dielectric rod with an extended probe feed acts as a field-launching element that excites the four-element CDRA. However, proper selection of the feeding element and optimum design elements offer an approximately 29% impedance bandwidth.

In [15], CDRA is excited through a dual-feeding network for achieving a wideband frequency with low cross-polarization and directive gain. Here, a dual differential probe feed network arrangement for CDRA is accomplished. This proposed structure offers a large-bandwidth, directive gain with low cross-polarization. More than 68% operational bandwidth (S_{11} < -10 dB) is achieved. The CDRA is fabricated from $\varepsilon_r = 12$ dielectric material, and dual SMA connector coaxial probes are connected to the extended wire, which is inserted in the DRA for excitation. This feed arrangement generates the same amplitude

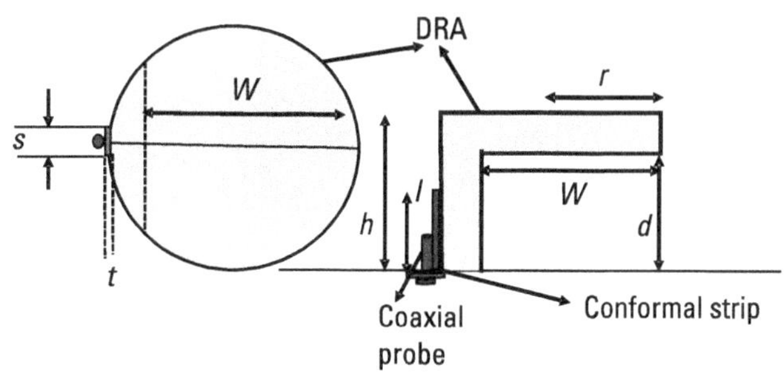

Figure 2.11 Top and side views of antenna [11].

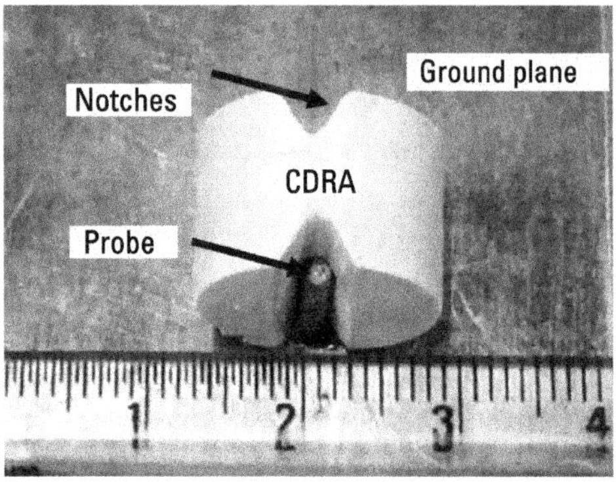

Figure 2.12 Fabricated structure of a laterally placed CDRA with triangular notches [12].

with 180° phase difference of excitation fields in DRA. Due to this, multiple resonances with a wide frequency band are achieved. Figure 2.13 represents the dual-feeding arrangements of the DRA. HDRAs geometries are examined in [16–18] for wideband applications. In [16], an inverted truncated hemispherical cup-shaped DRA is explored to obtain an 83% impedance bandwidth. The TM_{101} mode of an HDRA and the $TM_{01\delta}$-like mode of a CDRA are generated by the proper selection of the feed length (l = 1.2 cm) and the position of the coaxial probe. At a 2.3-GHz frequency, a 4.5-dBi gain is achieved. Figure 2.14 shows the fabricated prototype structure; an SMA connector is used to provide coaxial feeding to the fabricated prototype.

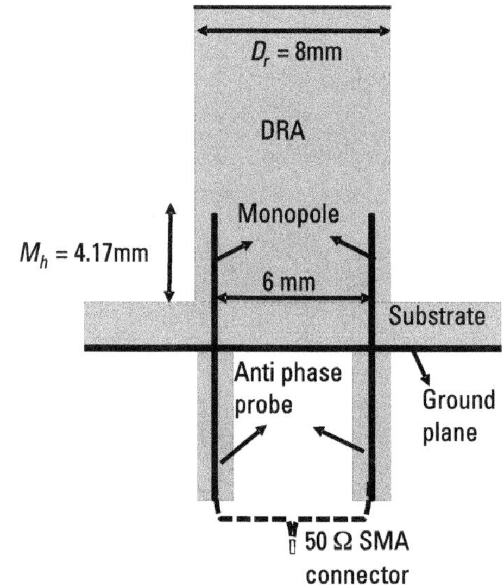

Figure 2.13 DRA excited by a differential dual-feeding coaxial probe [15].

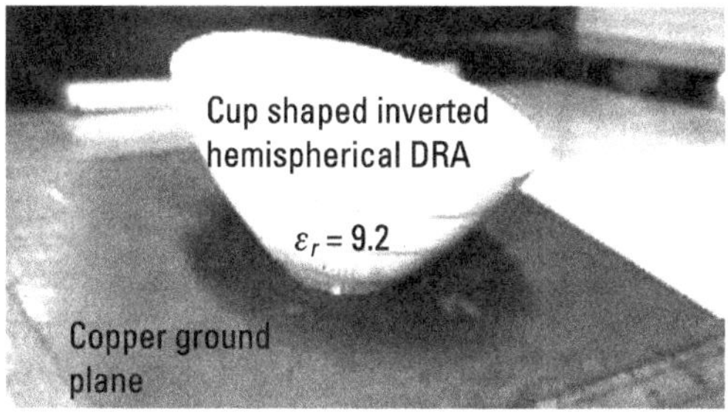

Figure 2.14 The fabricated prototype structure [16].

Furthermore, in [17], a conical shaped geometry is examined to obtain a 50% impedance bandwidth. A split cone is placed on the ground plane while a coaxial feeding network generates various modes. These modes are merged and offer the widest frequency response. Figure 2.15 illustrates the split-cone DRA configuration. To design a wide-bandwidth DRA, a truncated tetrahedron DRA is excited by a coaxial probe. This is investigated in [18]. The antenna offers a wide bandwidth when the small section of the cone is connected to the ground plane.

2.1.2 Microstrip Feeding

For integration with the printed circuit board (PCB), a monolithic microwave integrated circuit (MMIC) fabrication, microstrip feed excitation technique is very effective to excite the DRA. Microstrip feeding excites various modes in the DRA. The $HE_{11\delta}$ radiating mode is excited in the CDRA [19], as shown in Figure 2.16.

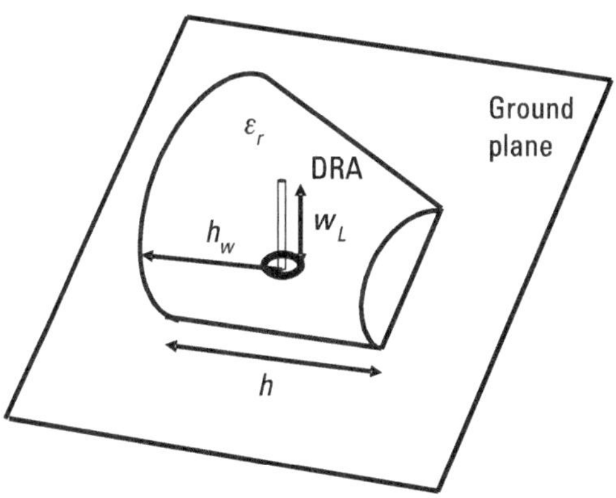

Figure 2.15 Split-cone DRA configuration [18].

Through the excitation, the amount of coupling from the microstrip feed line to the DRA can be controlled by the width and length of the microstrip feed line. For good radiation efficiency, the microstrip feed line width (W) can be calculated by

$$W = \frac{c}{2f_r}\sqrt{\frac{2}{\varepsilon_r + 1}} \tag{2.17}$$

where c is the velocity of light, f_r is the resonant frequency, and W is the width of the microstrip feed line. Also the amount of coupling depends on the dielectric material used to fabricate the DRA. A DRA composed of a high-permittivity dielectric material provides a strong coupling, while a DRA composed of a low-permittivity dielectric material facilitates a smaller amount of coupling. The microstrip feed line length can be adjusted across the outside of the DRA, which is known as the conformal microstrip feed line. Conformal and simple microstrip feed lines are used to propagate multiple resonant modes to achieve a wideband response. Meanwhile, it is quite challenging to achieve a wideband frequency response by providing excitation through a microstrip feed line to lower the dielectric constant material of the DRA. A copper tape is used during fabrication to accommodate the width and length of the conformal microstrip coupling line. While the DRA is fixed on the microstrip line, a small undesired air gap gets incorporated and reduces the effective dielectric constant value of the DRA. In a practical case, the conformal stretch of the line needs to be shorted to provide effective conductivity as well. The design equation of a microstrip line reveals that it is a band-limited transmission line section. Thus, to achieve a wideband response, an impedance transformer concept is deployed. For a very wideband response, tapered lines can also be used.

Various DRA geometries are excited through microstrip feed techniques and explored in [20–29]. A rectangular DRA is excited by a differential

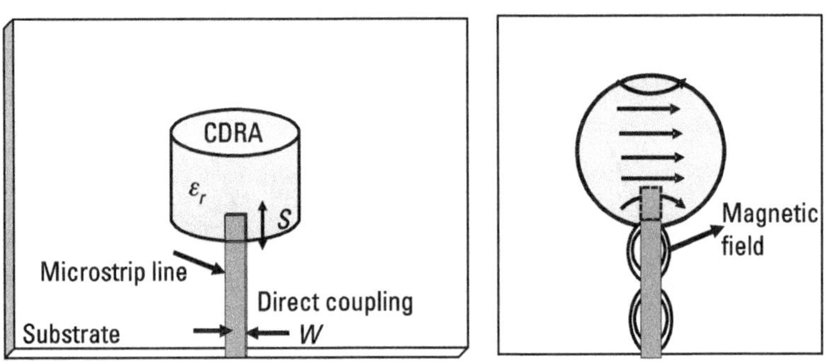

Figure 2.16 Microstrip feed line and magnetic fields across the feed line [19].

microstrip feed line in [20]. The TE_{111} mode is generated at a 2.4-GHz frequency, and a 10.4% impedance bandwidth is achieved due to phase differences arising by feeding excitation. A differentially fed RDRA is constructed by $\varepsilon_r = 10$ permittivity of dielectric material, and two feeding strips are incorporated by adhesive conducting tape. A 3-dB hybrid coupler is used to generate a differential feeding signal in the DRA. Figure 2.17 shows the excitation of a differential feed network for the RDRA. Moreover, [21] shows that bandwidth enhancement with miniaturization can be achieved by using a high aspect ratio two segmented DRA excited by microstrip feed. The high aspect ratio and permittivity of the antenna is used to generate two appropriate resonant modes, which are merged for obtaining an 11% impedance bandwidth. The proposed DRA is placed on top of an open-circuit microstrip feed line, and it is designed for a 50-Ω characteristic impedance. To achieve maximum coupling, the gap between the microstrip feed line and the DRA is crucial. A conformal-strip-excited asymmetric RDRA with a defected ground is investigated for ultra-wideband application in [22]. The antenna geometry covers a 124.32% (2.8–12 GHz) impedance bandwidth and excites numerous higher-order modes.

In [23], the authors report the use of a microstrip-coupled CDRA for enhancement of the impedance bandwidth. The effect of the energy coupling and air gap between the microstrip feed line to the CDRA is studied to improve the diversity as well as the bandwidth. The antenna was fabricated with $\varepsilon_r = 37$ dielectric material and excited by a 50-Ω open-ended microstrip line for impedance matching. Through an energy-coupling microstrip feed line, the $HE_{11\delta}$ mode is excited. Figure 2.18 depicts the top and side views of the antenna.

In [24], a half CDRA excited by a microstrip feed line is investigated for 2.4-GHz WLAN application. A new microstrip feeding technique is explored for a CDRA to obtain a balanced coupling mechanism, through a modified T-shaped and L-shaped microstrip feed line [25]. This antenna offers a 4–11.736-GHz (58.7%) wide frequency band. After adding a C-shaped strip line between the transmission line and the CDRA, the energy coupling is improved, thus leading to improvement in the gain of the antenna.

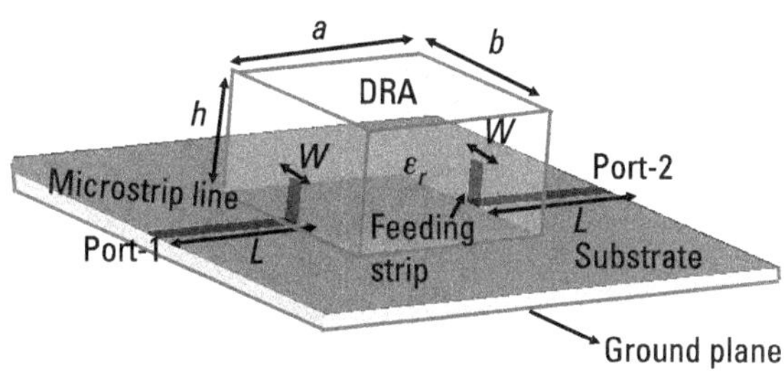

Figure 2.17 Excitation of a differential feeding network by an RDRA [20].

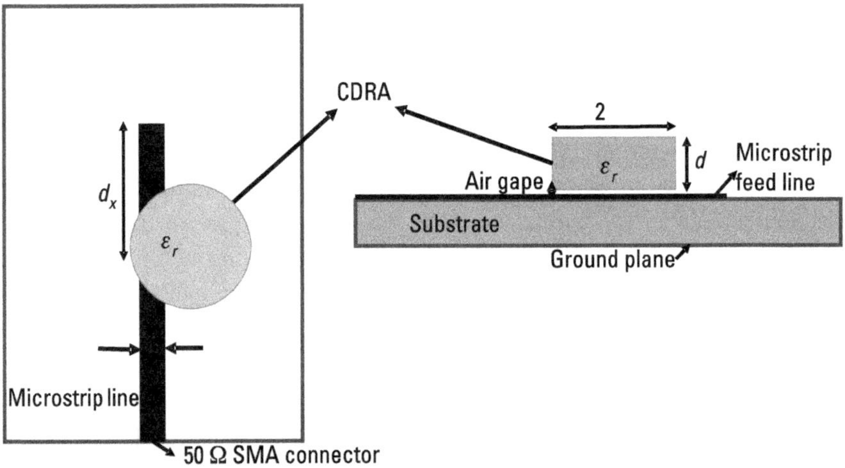

Figure 2.18 Top and side views of a CDRA [23].

In [26], a coplanar parasitic microstrip feed technique is investigated for linear polarization and the $HE_{11\delta}$ radiating mode. Also, for achieving wideband, an inverted half CDRA excited by a microstrip feed line was investigated. Interestingly, the author determined that the overall radiating surface area of antenna is increased. The antenna volume-to-surface area ratio decreases, and thus the Q factor of the antenna decreases and the overall bandwidth of the antenna increases. Figure 2.19 shows the geometry of the inverted half CDRA.

In [27], a compact CDRA with a circular slot is investigated for tri- band wireless application. The CDRA is excited by a modified angular-shaped microstrip feed line, and the use of a two ring–shaped resonator on the opposite side of the feed line attenuates unwanted frequency band. In the fabrication of this antenna, alumina ($\varepsilon_r = 9.8$) material is used. This antenna offers 2.35–2.55-GHz, 3.35–3.65-GHz, and 5.1–5.42-GHz frequency bands with a 4.71-dB average gain, respectively, and it is suitable for WLAN and WiMAX applications.

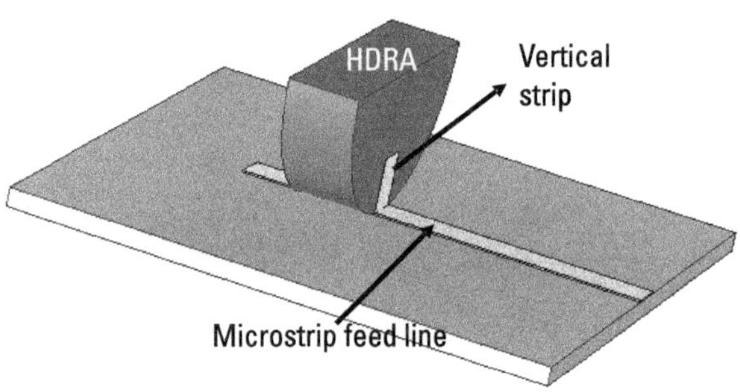

Figure 2.19 Geometry of an inverted half CDRA [26].

Similarly for quad-band cylindrical DRA, an annular-shaped microstrip with a complementary C-shaped etched ground plane is proposed [28]. Adjustment in the microstrip feed line is done to get the maximum coupling. When the microstrip feed line is placed at the center or shifted from the center position, $HE_{12\delta}$ and $HE_{11\delta}$ radiating modes are generated at 12.1 GHz and 11.6 GHz, respectively. The proposed antenna geometry offers 3.21–3.63-GHz (14.36%), 4.83–6.78-GHz (23.14%), 7.25–7.77-GHz (3.68%), and 11.76–12.45-GHz (1.88%) wideband frequencies, respectively. This antenna design is convenient for WLAN- (5.15–5.35/5.725–5.825-GHz) and WiMAX- (3.400–3.69/5.250-5.850-GHz) like applications. Figure 2.20 shows the fabricated antenna structure.

To achieve the triple operating band, a vertical and phi-shaped microstrip feed line is used to excite the CDRA [29]. A vertical microstrip line generates the $HE_{11\delta}$ mode, and a vertical phi-shaped line generates the $TM_{01\delta}$, $HE_{12\delta}$ mode, respectively. Hence the $TM_{01\delta}$ and $HE_{12\delta}$ radiating modes act like vertical and horizontal electric dipoles, and the $HEM_{11\delta}$ mode behaves as a horizontally placed magnetic dipole.

2.1.3 Aperture or Slot Feeding

A slot aperture with a microstrip transmission line is one of the most effective approaches to excite an antenna and integrate with a PCB/printed design technology. In this technique slots are created on the ground plane. As well, the DRA is fixed on the aperture or slot area, and energy coupling occurs between the microstrip feed line and the DRA through this slot area. Therefore, the radiating amount of energy below the ground plane can be controlled by the slot dimensions. Various types of apertures, including C-shaped, T-shaped, and annular ring–shaped, and cross-shaped slots are used to excite the CDRA [30–34]. The aperture or slot-coupling feeding technique provides the isolation between the microstrip transmission line and the DRA. Due to presence of the microstrip line below the ground, unwanted spurious radiation is reduced, which offers a significant advantage for the antenna radiation. Figure 2.21 shows the various slot or aperture geometries. The aperture microstrip

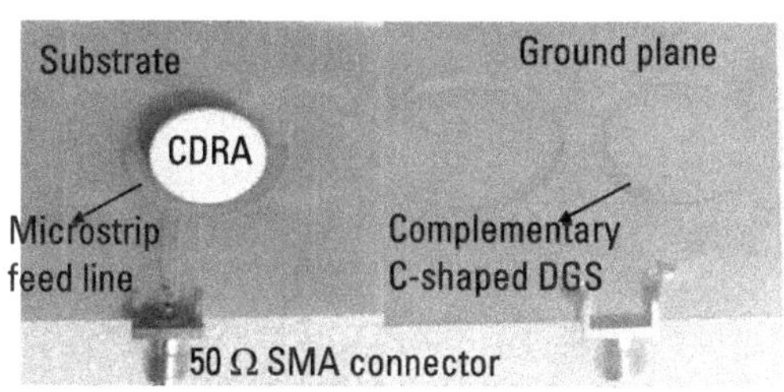

Figure 2.20 Fabricated antenna structure [28].

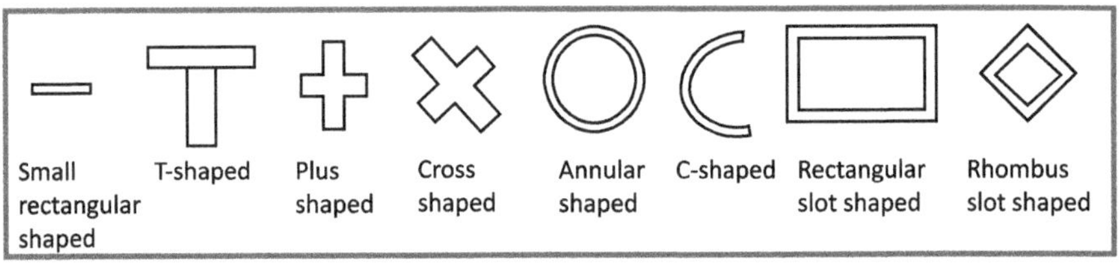

Figure 2.21 Various aperture and slot geometries.

transmission feed line width (w) and length (l) can control the coupling of energy as in the microstrip feeding technique.

The slot dimension length (l_s) and width (w_s) can also control the amount of coupling from the DRA to the microstrip transmission line. Also, according to the image theory concept, the flow of the magnetic currents should be double at aperture slots. Overall the aperture slot's surface integral of magnetic field can be calculated by the following [33]:

$$H_{tan}^{dra}\left(J_{dra}\right)+H_{tan}^{dra}\left(M_{dra}\right)-2H_{tan}^{dra}\left(M_{s}\right)=H_{tan}^{l}\left(M_{s}\right)+H_{tan}^{inc}$$

on closed surface s (2.18) where H_{tan}^{dra} is the tangent magnetic field within the DRA, J_{dra} is the equivalent electric surface current within the DRA, M_{dra} is the equivalent magnetic surface current within the DRA, Ms represents the equivalent magnetic currents of the aperture slots, H_{tan}^{l} is the tangent magnetic field within the substrate, and H_{tan}^{inc} is the impressed magnetic field. Figure 2.22 represents the aperture slot microstrip feed excitation and the flow of the equivalent magnetic current.

Strong energy coupling is required to achieve the wideband frequency response, so aperture slots are created in either the strong E field or the strong H field region of the ground plane. Figure 2.23 shows the coupling through the slots aperture. Some degree of antenna impedance matching can be achieved by

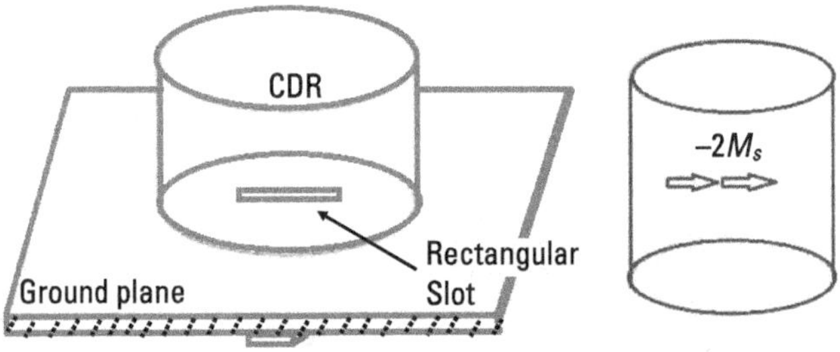

Figure 2.22 Aperture slot microstrip feed excitation and the flow direction of the equivalent magnetic current. The flow of the magnetic current doubles at the aperture slots based on the image theory concept [33].

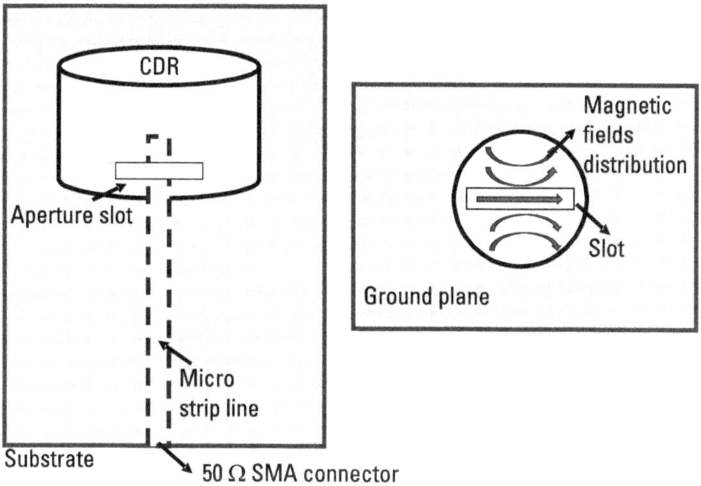

Figure 2.23 Slot aperture coupling through the CDRA and magnetic field distribution ($HE_{11\delta}$) [34].

the DRA by either placing the DRA at the center of the slot or offsetting it from the slot. The $HE_{11\delta}$ radiating mode is generated in the CDRA when excited through the rectangular slot [34].

The impedance of the antenna can be improved by adjusting the extension length of the microstrip line when the microstrip transmission feed line is used in the aperture-feeding technique. The extended portion of the stub of the transmission line behaves like a reactive component, and it reduces aperture admittance. Similarly, in the microstrip patch antenna, extended stub length has to be taken to be equivalent to $S_g = {\lambda_g}/{4}$, and λ_g is a guided wavelength of microstrip transmission line. To avoid excessive radiation through the slots, small area aperture slots have to be kept. Several aperture slot-feeding techniques have been reported by the authors for achieving the wideband frequency response with various radiating modes and radiation patterns [35–45].

In [35], to obtain wider bandwidth, a slot antenna with DRA is used. The slot dimensions are intended to be resonant at a predetermined frequency and then the natural resonant frequency of the dielectric structure is merged to achieve a wide bandwidth. The dielectric resonator is coupled to a microstrip-fed slot in the ground plane as shown in Figure 2.24. The size and position of the slot are optimized in order to enhance the impedance matching. Slot resonance happens at $\lambda_g/2$; here λ_g is wavelength inside the dielectric, which depends on the effective dielectric permittivity. The slot can be modeled by a magnetic current source, $\vec{M} = -\hat{z} \times E\hat{y}, E\hat{y}$ is the electric field in the slot induced by currents on the ground plane, and the magnetic current source lies in the direction along the bottom of the DRA. Another technique in [36] is explored

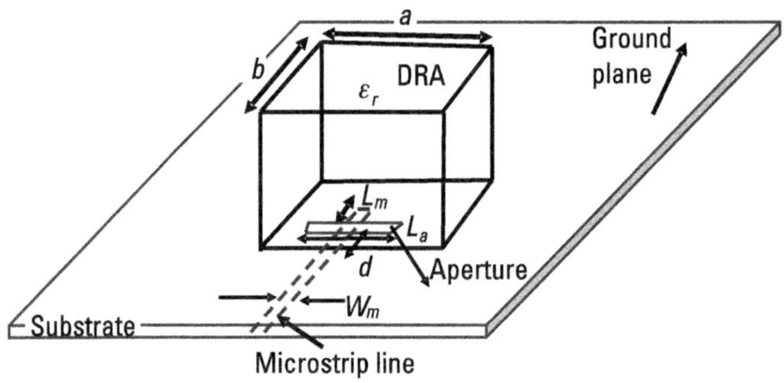

Figure 2.24 Dielectric resonator coupled with a microstrip-fed slot [35].

to achieve dual-band for ISM (2.4 GHz) and Wi-MAX (3.3–3.6 GHz) band applications. RDRA, modified cross shaped with partial ground plane generates orthogonal modes. This antenna is constructed from alumina (Al_2O_3) dielectric material having a dielectric permittivity $\varepsilon_r = 9.8$. The antenna covers an 11.78% (2.32–2.61-GHz) and 9.97% (3.24–3.58-GHz) operating frequency range.

In [37], the effect of aperture slot size and air gap between CDRAs to slots is studied on the resonant frequency, directivity, unloaded Q factor, and impedance bandwidth. Here, the aperture slot length (L_a) and width (W_a) couples the energy from the microstrip transmission line to the CDRA. The microstrip feed line is etched on the bottom side of the substrate, and the aperture slot is created on the top side of the ground plane substrate. For impedance matching, the open stub microstrip line is extended from the center to $L_p = 8$ mm. When excitation is given to the CDRA after placing it at the center of the aperture slots, the strongest magnetic field coupling is obtained and a broad side radiation pattern is achieved. An $HE_{11\delta}$-like radiating mode is generated in CDRA. For maximum energy coupling, the aperture slot dimension and air gap effects are optimized. Figure 2.25 shows the bandwidth versus aperture slot length (L_a) for various slot widths (W_a).

In [38], a simple CDRA excited by an aperture-coupled microstrip transmission feed line was investigated for wideband wireless applications. To improve the bandwidth, a fork-like stub was added with a 50-Ω microstrip line. It offers two resonant modes for wide impedance bandwidth. Figure 2.26 depicts an aperture slot-coupled feed CDRA with a fork-like tuning stub. The CDRA was designed for the $HE_{11\delta}$ radiating mode, and its resonant frequency is calculated based on the dimensions (a and d) and dielectric constant (ε_r) of the dielectric resonator using (2.19) [38]. Here c is the speed of light, and a, d, and ε_r are, respectively, the radius, height, and relative dielectric constant of the dielectric resonator. Additionally this antenna offers a 7.9-dBi gain.

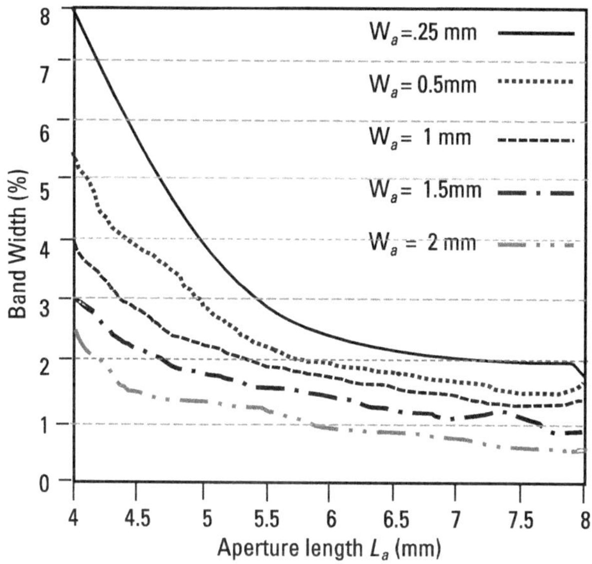

Figure 2.25 Bandwidth versus aperture slot length (La) for various slot widths (Wa) [37].

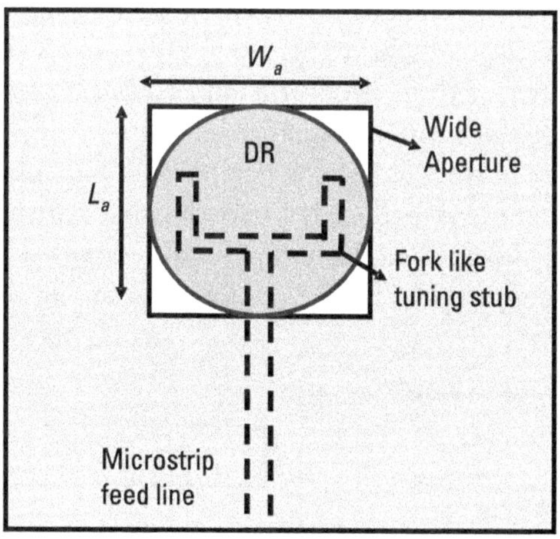

Figure 2.26 Aperture slot–coupled feed CDRA with a fork-like tuning stub [38].

$$f_r = \frac{6.324 \times c}{2\pi a\sqrt{\varepsilon_r + 2}}\left(0.27 + 0.36\frac{a}{2d} + \left(0.02\frac{a}{2d}\right)^2\right) \tag{2.19}$$

Similarly in [39], the author reports an asymmetric aperture-coupled microstrip feeding to a dielectric resonator for wideband applications. Here two cylindrical dielectric resonators are placed on the asymmetric side with respect to the center of the rectangular aperture slot on the ground plane. The antenna's resonant frequency and the amount of undesired radiation can be controlled by the aperture slot dimensions. This antenna covers the 9.62–12.9-GHz (29%) frequency band with an 8-dBi maximum gain. For wideband applications, an extended offset aperture-coupled double-cylindrical DRA is investigated in [40]. Two CDRAs are placed on a ground plane asymmetrically with respect to the center of offset rectangular aperture slots. The effect of the rectangular slots on return loss is studied. The combined effects of CDRA radiation modes and slot radiation modes are investigated, and $HEM_{01\delta}$-, $TE_{01\delta}$-, $HE_{11\delta}$-, $EH_{11\delta}$-, and $TE_{01\delta}$-like modes are obtained. The antenna offers a 5.9–7.32-GHz frequency band in the lower sideband and a 8.72–15-GHz frequency band in the upper sideband with a 12-dBi maximum gain. Figure 2.27 shows the geometry of the aperture slot with the CDRA, the field distribution of individual slots, and the equivalent merged slots.

A CDRA designed for dual-band and wideband dual polarization is explained in [41]. Antenna dual-port excitation and strip- and slot-feeding excitation techniques are used to generate dual polarization and wide bandwidth. This antenna operates for distributed control system (DCS) (1.7 –1.88 GHz) and wireless local area network (WLAN) (2. –2.48 GHz) bands, respectively. Due to dual-feed excitation, fundamental HEM_{111} and higher-order HEM_{113} are excited in DRA. In [42], a hybrid-aperture CDRA with a dual-band and dual polarization is investigated for wideband applications. For antenna excitation, a pentagon aperture slot contributes in generating $HEM_{11\delta}$ and $HEM_{12\delta}$ radiation hybrid modes. Further, adding a quarter annular stub with a microstrip transmission line creates a path delay between the two modes, thus achieving wideband and circular polarization. This antenna is suitable for WiMAX (2.5/5.5-GHz) and WLAN (2.5/5.5-GHz) wireless applications.

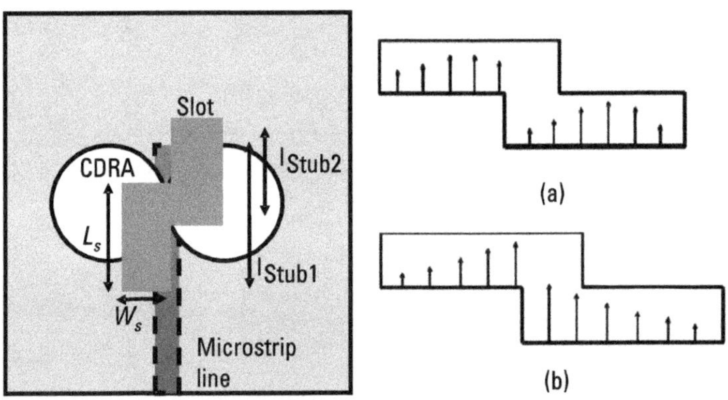

Figure 2.27 Aperture slot with CDRA and field distribution of (a) individual slots and (b) equivalent merged slots [40].

In [43], an HDRA is excited through aperture coupling. The effects of the slot's length, the slot's position, and the slot's width on the broadside TE_{111} mode input impedance are explored. The best impedance bandwidth (11.1%) is obtained when the slot length is $L = 11$ mm. Figure 2.28 shows the aperture-coupled HDRA.

2.2 Bandwidth Enhanced by Stacking of Various Permittivity Layers

This section investigates the stacking of various dielectric materials for bandwidth improvement. Figure 2.29 depicts the geometry of stacking dielectric material over CDRA [44]. Each dielectric material has a different resonant frequency, and these resonant frequencies combine, thus offering a wide frequency range of operation. Figure 2.30 shows the frequency response of the stacked dielectric material CDRA. If two dielectric materials have resonant frequencies f_{r1} and f_{r2} and bandwidths $B.W_1$ and $B.W_2$, then the overall combined bandwidth response may be larger than that of $B.W_1 + B.W_2$ (i.e., overall $B.W. \geq B.W_1 + B.W_2$). It provides a wide frequency band for wireless applications when the dielectric material is properly selected and the resonant frequency is properly tuned in the antenna. The number of modes of stacked DRA depends on the excitation technique, feed location, and feed dimensions.

In Figure 2.31, n is the number of stacked dielectric materials in the CDRA. The total effective dielectric constant by stacking of n numbers of dielectric material of any DRA can be calculated by the assumption that it is the weighted average of the overall dielectric material ε_{r1}, ε_{r2}, ε_{r3}...... and ε_{rn} as calculated in (2.20). This approximate modeling of the n numbers of dielectric material DRA structure is calculated by a simple static capacitance model concept [45]. Here in (2.20) ε_{r1}, ε_{r2}, ε_{r3} ...... ε_n are dielectric material constants and V_1, V_2, V_3...... V_n are the volumes of the dielectric material layer.

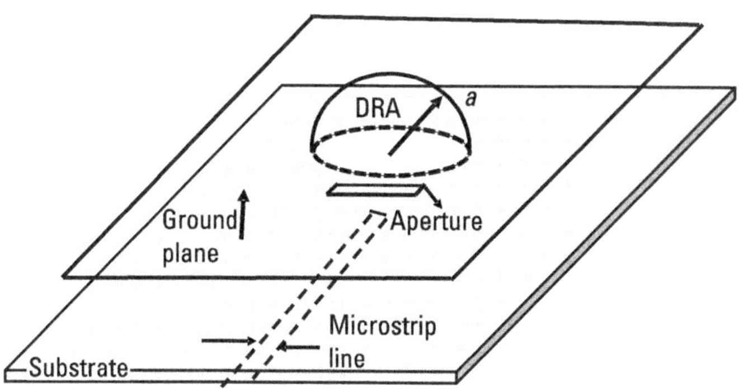

Figure 2.28 Aperture-coupled hemispherical DRA [43].

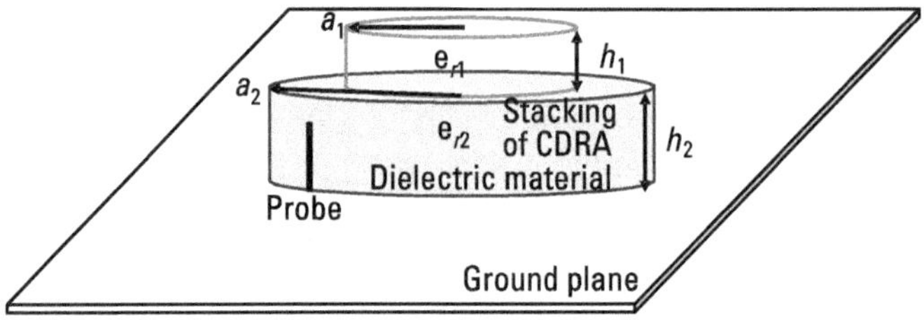

Figure 2.29 The geometry of stacking dielectric material over CDRA [44].

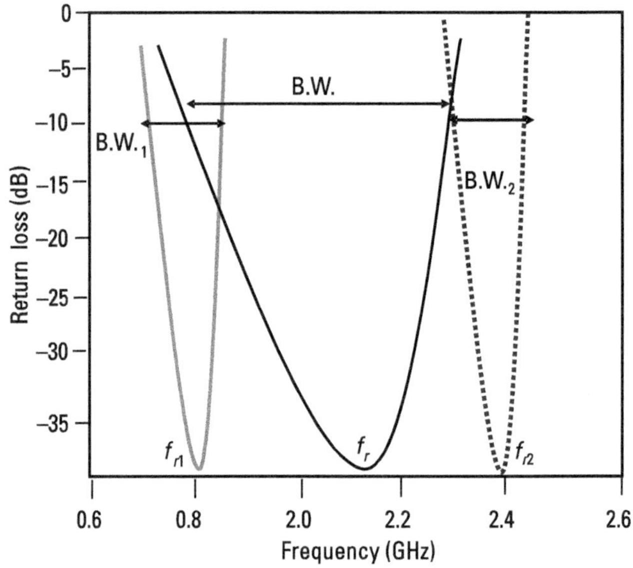

Figure 2.30 The frequency response of stacking dielectric material over CDRA.

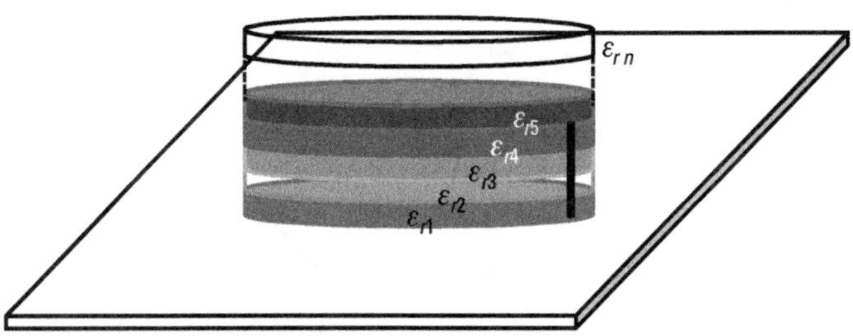

Figure 2.31 *N* numbers of dielectric layers in CDRA.

$$\varepsilon_{eff} = \frac{\varepsilon_{r1} \times V_1 + \varepsilon_{r2} \times V_2 + \varepsilon_{r3} \times V_3 + \ldots + \varepsilon_{rn} \times V_n}{V_1 + V_2 + V_3 + \ldots + V_n} \tag{2.20}$$

In addition, the resonant frequency of DRA is dependent on an effective dielectric constant value. If the total calculated effective dielectric material constant reduces, the quality factor also reduces, thus increasing the overall impedance bandwidth of the antenna [46]. The DRA quality factor and bandwidth are calculated from (2.21) and (2.22). The Q factor and bandwidth totally depend on the effective dielectric constant of the DRA.

$$Q \propto 2\omega_0 \left(\varepsilon_{\text{eff}}\right)^p \left(\frac{volume}{surface}\right)^s \tag{2.21}$$

$$B.W. = \frac{VSWR - 1}{Q\sqrt{VSWR}} \tag{2.22}$$

We now consider, for bandwidth improvement, techniques for the dual-layer, triple-layer, and multilayer stacking of dielectric material [47–51]. The stacking of different materials used in designing RDRAs is investigated in [47–55]. A two-segmented RDRA (multielement approach) is constructed by low-permittivity ($\varepsilon_l = 2.2$) and high-permittivity ($\varepsilon_u = 7.2$) dielectric material [48], which obtains a 30% impedance bandwidth. Similarly in [49], two-segment RDRAs placed side by side with different dielectric material ($\varepsilon_1 = 4.3$ and $\varepsilon_2 = 9.2$) are excited by microstrip feed. A further 2.8–13.49-GHz (131.24%) wideband frequency response is achieved when the rectangular slots are incorporated in the ground plane. The antenna is fabricated and achieves a 7.2-dBi gain at 10 GHz along with more than 70% of radiation efficiency over the frequency range. Figure 2.32 shows a fabricated prototype of the antenna. The principle of multielement DRAs follows the same static capacitance model of stacked DRAs.

In [50], the authors investigate stacked, core-plugged, and embedded stacked cylindrical geometries for bandwidth improvement. (See Figure 2.33.) The best results are obtained when parameters $a = 12.7$ mm, height $d = 22.5$mm, $d_{\text{lower}} = 12.7$ mm, and $\varepsilon_{r\,\text{lower}} = 12$, $\varepsilon_{r\,\text{upper}} = 20$, $\varepsilon_{r\,\text{plug}} = 1$ for stacked, core-plugged, and embedded stacked cylindrical geometries, respectively; 68.1% embedded stacked, 50.7% core plugged embedded, 59.9% stacked, and 21.0% homogenous DRA structure impedance bandwidth are achieved, respectively.

In [51], the authors report on a dual-segment composite quarter-CDRA for wideband wireless applications. A coaxial probe is used to excite the $TM_{01\delta}$

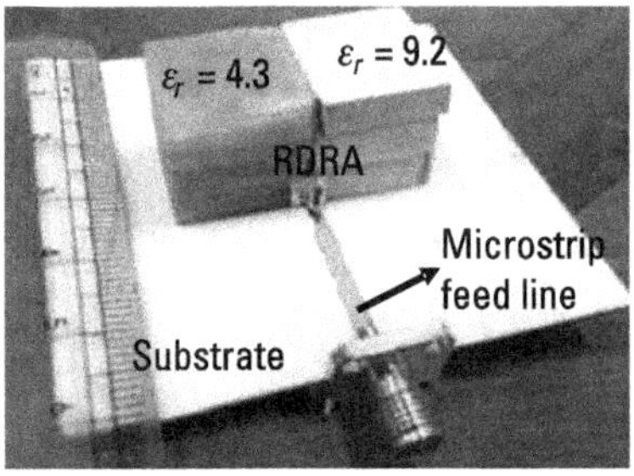

Figure 2.32 A fabricated prototype of an RDRA [49].

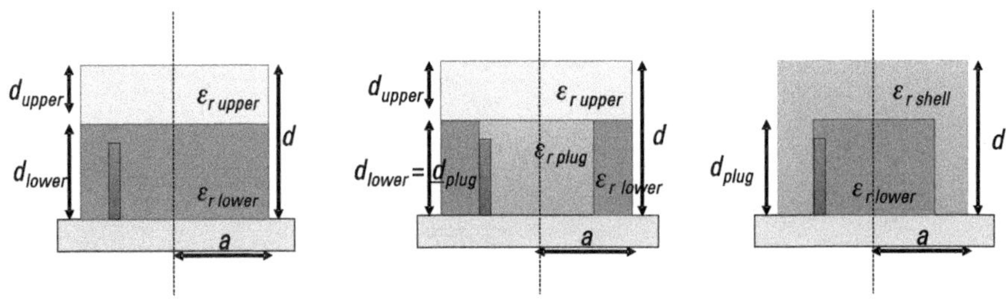

Figure 2.33 A side view of stacked, core-plug embedded, and embedded stacked CDRAs [50].

mode in a multisegment quarter-CDRA. A single-, dual-, and quarter-element, dual-segment composite CDRA is simulated and fabricated. Figure 2.34 shows these fabricated structures. An 85.13% impedance bandwidth is obtained by the four-element, dual-segment composite quarter-CDRA. This antenna covers C- and X-band wireless applications with a high gain. The four-element dual-segment composite quarter-CDRA upper ($\varepsilon_{r\,\text{upper}} = 9.8$) and lower ($\varepsilon_{r\,\text{lower}} = 2.1$) sections are the designs of Alumina Ceramic, Ants Ceramics Private Ltd., and Teflon, respectively. The effective dielectric constant of the lower and upper dielectric segments decrease, so the Q factor decreases, and overall bandwidth is improved. Table 2.1 lists the bandwidth performances of single-; dual-; and four-element, dual-segment composite quarter-CDRAs.

In [52, 53], a two-layer dielectric composite CDRA is considered for bandwidth improvement. Similarly, a four-element three-multilayer permittivity CDRA is presented in [54]. This antenna structure provides a 58.1% measured input impedance bandwidth at a 4.03-GHz resonant frequency also covering the 3.48 GHz–5.82-GHz frequency band. Figure 2.35 illustrates a four-element, three-multilayer permittivity CDRA structure. A half-split multilayer CDRA [55] reported a 60% impedance bandwidth with a 6.09-dBi maximum gain. Also, in [56], a three-layer concentric half-CDRA is reported for wide bandwidth. This antenna is designed with $\varepsilon_{r1} = 10.2$, $\varepsilon_{r2} = 6.15$, and ε_{r3}

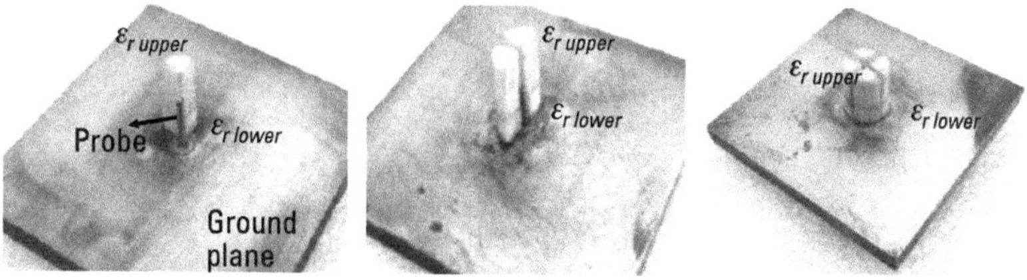

Figure 2.34 The fabricated structure of (a) dual-segment quarter-CDRA; (b) two-element, dual-segment composite quarter-CDRA; and (c) four-element, dual-segment composite quarter-CDRA [51].

Table. 2.1
Various Antenna Performance Statistics [51]

Parameter	Frequency Range		Resonant Frequency		Bandwidth (%)	
Antenna Geometry	**Simulated**	**Measured**	**Simulated**	**Measured**	**Simulated**	**Measured**
Dual-segment quarter-CDRA	7.5–13.5	7.0–13.0	9.5	9.45	63.15	63.49
Two-element, dual-segment composite quarter-CDRA	6.4–12.35	6.2–12.2	8.05	7.75	73.91	77.41
Four-element, dual-segment composite quarter-CDRA	5.1–11.5	5.1–11.4	7.5	7.45	85.33	85.13

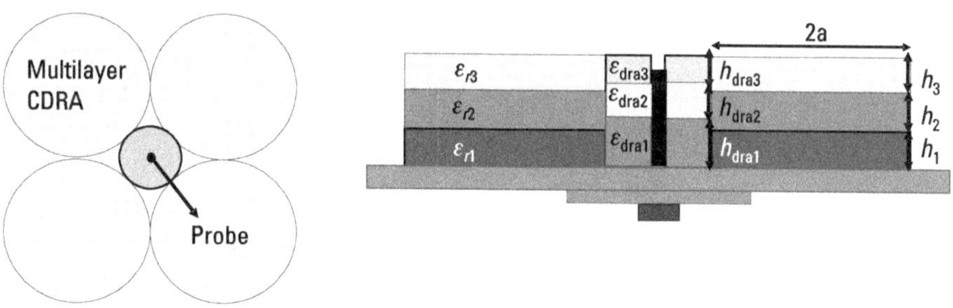

Figure 2.35 A four-element, three-multilayer permittivity CDRA structure: (a) top view and (b) side view [54].

= 2.32 dielectric materials. The antenna-optimized geometry offers an 82.01% impedance bandwidth and covers the frequency range from 4.51 to 9.17 GHz.

Furthermore, in [57], the variation of permittivity is studied in the radial direction of a concentric half-split CDRA for bandwidth enhancement.

Bandwidth enhancement techniques are investigated for homogeneous, two-layer, and three-layer permittivity composition in the CDRA. The thickness of each layer of dielectric constant is selected uniformly while designing a three-layer concentric half-split CDRA. The radii and dielectric material constants of the antenna are r_1, r_2, r_3, ε_{r1}, ε_{r2}, and ε_{r3}, respectively. For best results, the parametric study is done on the cylinder radius and various dielectric permittivity arrangements. Here, $r_1 = 5$ mm, $r_2 = 10$ mm, $r_3 = 15$ mm, $\varepsilon_{r1} = 10.2$, $\varepsilon_{r2} = 6.15$, $\varepsilon_{r3} = 2.32$, and $h = 11.4$ mm are optimized elements that have been used for designing the antenna. The fabricated antenna offers a 69.67% impedance bandwidth corresponding to a 6.84-GHz resonant frequency. The proposed configuration also attains an 8.94-dBi average gain for communication.

Next in [58], permittivity variation in the axial direction of the CDRA is reported for wideband with fabrication limitations. For the $HE_{11\delta}$ operating mode, a homogeneous, two-layer, three-layer, and four-layer multipermittivity CDRA are designed. Also the antenna performance, aspect ratio, dielectric layer arrangement, layer thickness, layer permittivity, and probe length are studied. A multilayer arrangement of CDRA offers a low effective dielectric constant value. Various dielectric permittivity layers have different operating frequencies and radiation Q factors; these multilayer arrangements can be used simultaneously to control the operating frequency as well as the radiation Q factor of the DRA. The radiation Q factor is dependent on the effective permittivity of the DRA, and the resonant frequency is inversely proportional to the effective permittivity of the antenna. Selecting a combination of low effective dielectric permittivity material can significantly improve the bandwidth. In [59], the author reports on a two-layer HDRA for wideband applications. It is excited by a coaxial probe and achieves an omnidirectional TM_{101} radiating mode. Acrylic and glass materials are used for the construction of the inner and outer transparent layers of the HDRA, respectively, and the light source consists of two LEDs that are inserted into the inner (acrylic) layer of the HDRA; a 31.9% impedance bandwidth is obtained by this antenna geometry. Figure 2.36 depicts the light source (two LEDs) and the omnidirectional transparent two-layer HDRA.

In [60], an HDRA is split into four uniform quarters and excited by a coaxial probe at the center of the ground plane for a wideband application. A four-element quarter-HDRA resonating with dominant and first higher modes corresponds to a 30% impedance bandwidth and offers a monopole-like radiation pattern.

2.3 Metallic Loading

Here we discuss DRAs with metallic part-loading hybrid antenna techniques (including monopole insertion, microstrip patch, and metallic patch-loading

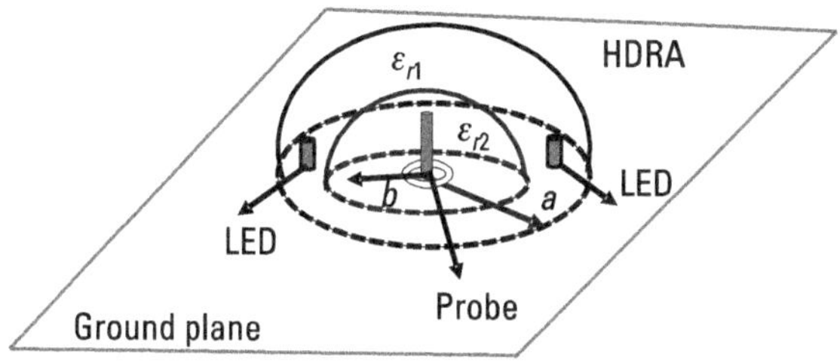

Figure 2.36 A two-layer HDRA [59].

to the top of a DRA), with an emphasis on bandwidth improvement. The bandwidth of a DRA can significantly improve with the loading of a monopole antenna, which is shown in Figure 2.37 [61]. The CDRA and monopole are placed at the center of the ground plane. The monopole acts as a quarter-wavelength radiator and can also be used to give the excitation to the DRA. For the $TM_{01\delta}$ mode, a CDRA is constructed with a near-field concentration similar to the monopole antenna radiation pattern [61, 62]. In this manner, the monopole can efficiently excite the DRA. The single-monopole antenna and DRA operate in different resonance frequencies. If the monopole and DRA are integrated in a single geometry then impedance matching is achieved at their resonant frequency, because their radiation modes are merged, offering a wide-band frequency range.

To construct a CDRA with a monopole-loaded antenna, various design parameters, such as dielectric material (ε_{dr}), aspect ratio (a/h), and monopole antenna length (l_m), can effectively control the impedance bandwidth for long-distance wireless communication. The following design procedure obtains a broad bandwidth of operation [62]:

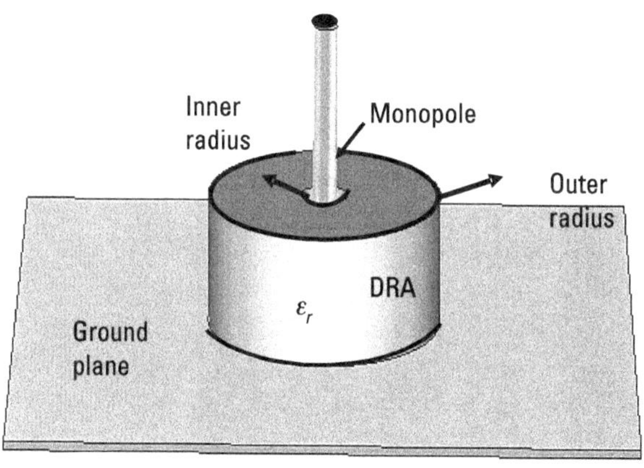

Figure 2.37 Monopole-loaded CDRA [61].

- The monopole and CDRA dimensions are calculated as follows:
  - The spacing(s) between the inner CDRA radius (a_i) to monopole is chosen as $0.13\lambda \leq s \leq 0.016\lambda$.
- For the desired frequency range, the monopole length is chosen to be a quarter-wavelength of lowest frequency (i.e., $l_m = \frac{\lambda}{4}$), where λ is the wavelength corresponding to the lowest frequency range. Similarly, the thin monopole wire radius has to be chosen from $0.004\lambda \leq r_m \leq 0.08\lambda$.
- For the $TM_{01\delta}$ radiation mode, the CDRA upper-side frequency (f_h) should be chosen to be 2.5 times the lowest side frequency (fl). Also the CDRA physical dimension is calculated by

$$a_i = r_m + s \tag{2.23}$$

$$a = {}^{a_i}\!/_{0.3} \tag{2.24}$$

- The height of the CDRA is such that $0.4\lambda \leq h_d \leq 0.5\lambda$, where a is the radius, and h_d is the height of the CDRA, and the CDRA dielectric constant can be calculated by

$$\varepsilon_{dr} = \frac{\left(\frac{\pi}{2h_d}\right)^2 + \left(\frac{4.71}{a}\right)^2}{\left(\frac{2\pi f_h}{c}\right)^2} \tag{2.25}$$

A microstrip patch is used with DRA hybrid techniques to expand the bandwidth [63, 64]. For bandwidth enhancement, the resonant frequency of the microstrip patch must be tuned with the DRA's resonant frequency. Figure 2.38 shows some hybrid antenna structures. This approach improves the impedance bandwidth of the conventional microstrip antenna by more than 10% [64]. In this hybrid method, the CDRA is mounted on the microstrip antenna. Here, the resonant frequency and exciting modes are controlled by the feed point of excitation. Impedance matching is obtained when the dielectric resonator is placed on the microstrip patch. Patch-loading on the top of the DRA reduces the lower sideband frequency and achieves compactness of the antenna structure without changing the physical shape of antenna. In [65–75], the authors report several research papers on metal integration with CDRA, which is dedicated to bandwidth expansion.

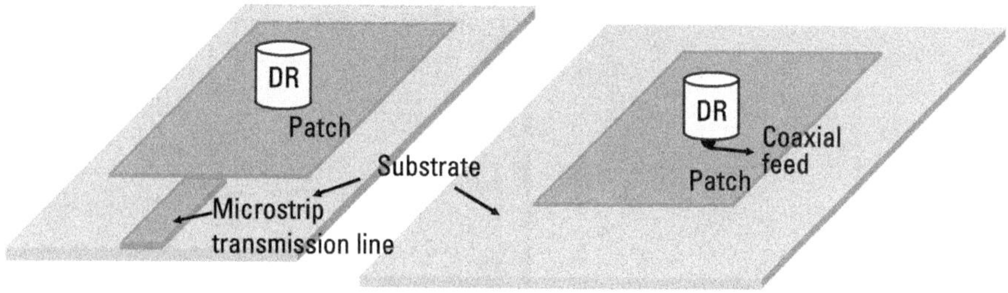

Figure 2.38 Hybrid antenna structure [63, 64].

A monopole-integrated dielectric resonator hybrid antenna is presented in [65]. This antenna is designed for various ultra-wideband wireless applications and covers the 4.5–11-GHz frequency band. The DRA aspect ratio (h/a) holds the important role of controlling resonance in this antenna design, and the loading effect excites the $TM_{01\delta}$ mode. To achieve a wideband response, it is necessary to make the proper selection of the optimized dimensions of the monopole and DRA. Therefore, the monopole and CDRA impedance bandwidths are matched, and the resonant frequency is tuned and merged. The $TM_{01\delta}$ and other higher-order modes are excited in CDRA when electric coupling occurs from the monopole to the DRA. Figure 2.39 shows a higher-order E field vector inside the CDRA. For fabrication of the antenna $\varepsilon_r = 10$, dielectric material is used, and this structure is placed on a circular ground plane. A centrally located SMA 8004 A connector is mounted on the back side of the ground plane, which is used to excite the antenna.

In addition, [66] discusses the use of a monopole antenna loaded by a step-radius CDRA for achieving ultra-wideband [66]. Figure 2.40 shows the geometry of a step-radius monopole with a dielectric hybrid CDRA. A coaxial-fed extended copper wire acts as a monopole that is loaded on a step-radius CDRA. The antenna is simulated using commercially available software

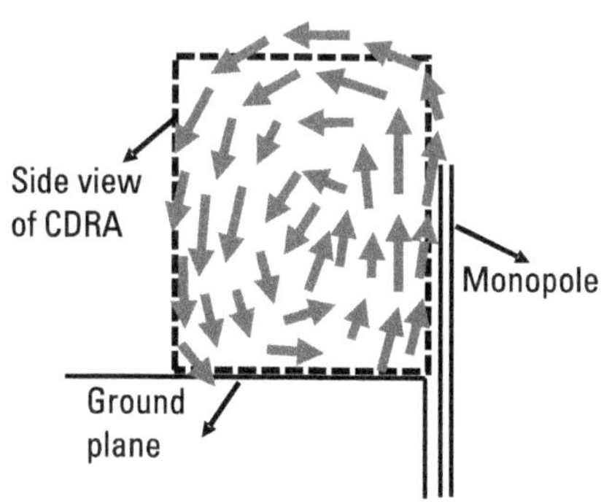

Figure 2.39 A higher-order E field vector inside the CDRA [65].

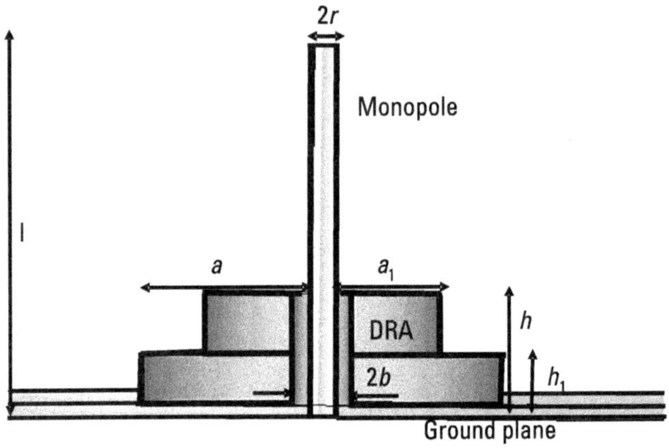

Figure 2.40 The geometry of a step-radius monopole with a dielectric hybrid CDRA [66].

such as computer simulation technology (CST) based on the finite-integration method (FIT) and high frequency structure simulator (HFSS) based on the finite-element method (FEM) for verifying the results. It offers a 3–10.3-GHz (110%) operating frequency range for numerous wireless applications. An extensive parametric study is applied on the dimensions of antennas for enhancing the bandwidth [66]. To achieve the ultra-wideband, first, monopole design parameters are selected for the lowest operating frequency, and CDRA design parameters are determined by

$$2r \leq b \leq 3r \tag{2.26a}$$

$$0.45l \leq h \leq 0.50l \tag{2.26b}$$

$$0.75h \leq a \leq 0.95h \tag{2.26c}$$

Here l and r represent the length and radius of the monopole, respectively; a, b, and h are the dimensions of the ring CDRA. Here wideband is achieved when impedance matching is obtained between the monopole and CDRA.

Similarly in [67], for bandwidth enhancement, a dielectric resonator is placed on a microstrip patch. The bandwidth of the microstrip antenna is doubled when the dielectric resonator is top-loaded. Another dielectric resonator on a patch antenna is investigated for broadband wireless applications in [68]. To enhance the bandwidth of the antenna, perfect matching of the exciting fields of the patch and dielectric resonator is required.

Subsequently, [69] investigates the loading of a circular metallic patch on a half-HDRA for wireless applications. The circular metallic patch plays an important role in the tuning of the frequency of antenna geometry. The radius of the patch controls the resonant frequency and bandwidth of operation. The metallic patch plays a significant role in improving the gain of the antenna up to 8.5 dBi. It acts as a reflector and hence improves the gain. An optimized radius of the patch ($r = 0.6$ cm) offers 0.82 GHz of impedance bandwidth. Figure 2.41 demonstrates a metal-loaded half-HDRA configuration.

In [70], a hemispherical and conical-shaped dielectric ring resonator with a monopole is explored for various wideband wireless applications. The spacing between the monopole surface and dielectric resonator is calculated by $s = b - r$. Here, b is a cut-out cylinder radius of the DRA, and r is the radius of the monopole. The centrally cut-out hole in dielectric resonator not only provides a space for accommodating the vertical monopole, but also plays a significant role in coupling the electromagnetic fields between the monopole and the dielectric resonant structure. Both DRA geometries integrated with a monopole offered a 126.5% impedance bandwidth. Figure 2.42 shows a hemispherical and conical-shaped dielectric ring resonator with a monopole DRA. An aperture-coupled multilayer hemispherical DRA with a conformal conducting patch is investigated in [71]. The electric surface current on the conformal patches are calculated from $J_s = n \times H$, where n is the unit normal vector to the surface of the spherical dielectric resonator. The antenna impedance is affected through the loading of metallic patches.

2.4 Compact and Low-Profile Geometry

Recent advances in wireless communication technology and the miniaturization of electronic devices have increased the need for compact device applications, including small antennas. A compact and low-profile DRA structure attracts a lot of attention for modern wireless technology applications like embedded PCB design, real-time weather monitoring systems, and the global system for

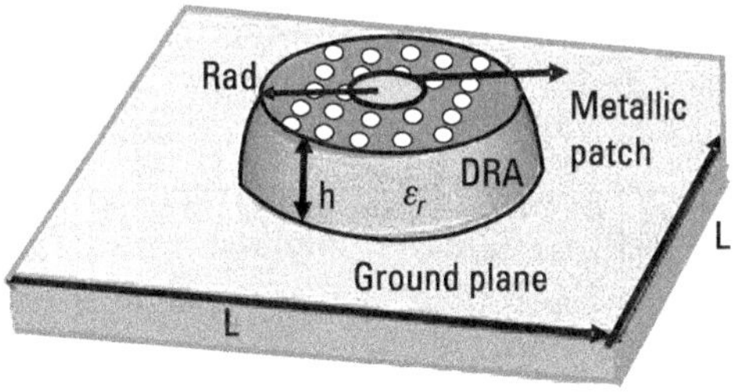

Figure 2.41 Metallic patch-loaded HDRA configuration [69].

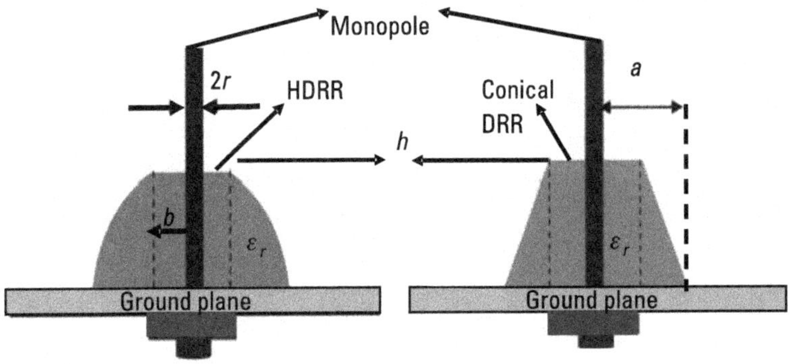

Figure 2.42 Hemispherical and conical-shaped DRA with a monopole DRA.

mobile communication [72–75]. It is always challenging to maintain the bandwidth and gain performance of a compact and low-profile antenna for the microwave frequency range. An antenna operates at a high-frequency range when the physical size of the antenna is reduced. To achieve compact antenna design in CDRA, the loading of metallic plates, posts, and strips is used to reduce the electric size of the antenna. Similarly, to make a low-profile antenna geometry, a high dielectric constant material is used to reduce the Q factor and improve the bandwidth of antenna. In addition, the height of a low-profile antenna must be $h \leq 0.03\lambda_0$. See [76–80] for further discussion of design techniques for compact and low-profile antennas that achieve a wide bandwidth.

A compact RDRA achieved 60–110% ultra-wideband in [72]. This DRA obtains multiple low-Q factor radiating modes with contiguous wide bandwidths. For antenna designing, a low-permittivity material ($\varepsilon_r = 2.2$) is inserted between high-permittivity ($\varepsilon_r = 9.2$) dielectric material DRA to the ground plane, and a planar conducting wall is applied to enhance the bandwidth performance as well as reduce the size of the antenna. The antenna is designed to cover the FCC UWB band from 3.1 GHz to 10.6 GHz, and it has a dielectric volume of $12 \times 8 \times 15.2$ mm^3. Figure 2.43 illustrates a half-size DRA with a finite conductor on one face.

In [73], a half-CDRA is studied for the miniaturization of the electric size of the antenna for broadband wireless application. For the design, an antenna metal plate is employed on the planar side of the half-CDRA. The volume of the CDRA is approximately diminished by half. Applying a metal plate perpendicular to the copper ground plane makes it act as a shorting post for the E field and allows the CDRA to be diminished to a symmetric axis. This shorting metal plate works like a shorting post that reduces the patch antenna length from $\frac{\lambda}{2}$ to $\frac{\lambda}{4}$; thus the metal plate acts as an electric wall, thereby achieving a

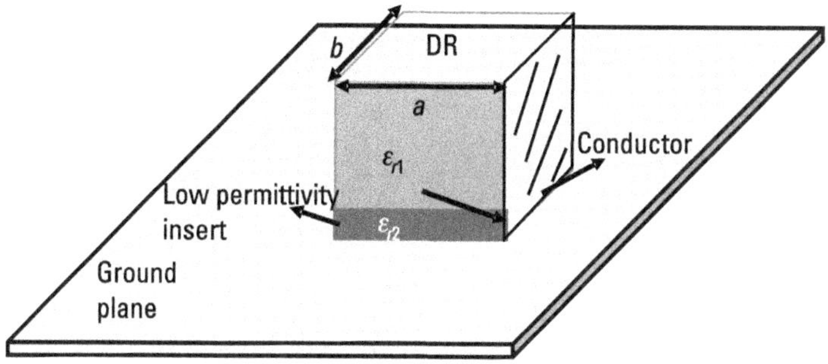

Figure 2.43 A half-size DRA with a finite conductor on one face.

miniaturized electric length. Figure 2.44 shows the geometry of a metal plate with a half-CDRA.

In [74], a compact cylindrical-sector DRA is investigated for bandwidth enhancement. The volume of the CDRA is reduced by removing sectors of dielectric material. Removing the arc through the sectored angle β and the metallization of the inner surface reduces the volume of the antenna with miniaturization of electric length. In addition, [75] proposes a compact CDRA for wideband wireless sensor network applications. The design of the antenna uses a stacked CDRA with a metallic patch on top of the CDRA. Figure 2.45 shows the geometry of the stacked antenna and its fabricated structure. Two important techniques are used for bandwidth and gain improvement, the stacking of two different dielectric materials and metallic patch-loading on top of the CDRA. This antenna covers the 5.7–13.2-GHz frequency range, and a $TM_{21\delta}$-like higher-order mode is excited. The stacking of two dielectric materials offers wide bandwidth, but the further addition of a metallic patch on top of the DRA helps to achieve the miniaturization of the electric length of antenna. Moreover, the antenna bandwidth improves when proper tuning is obtained between the resonant frequency of the CDRA and metallic patch. FR-4 (ε_r=

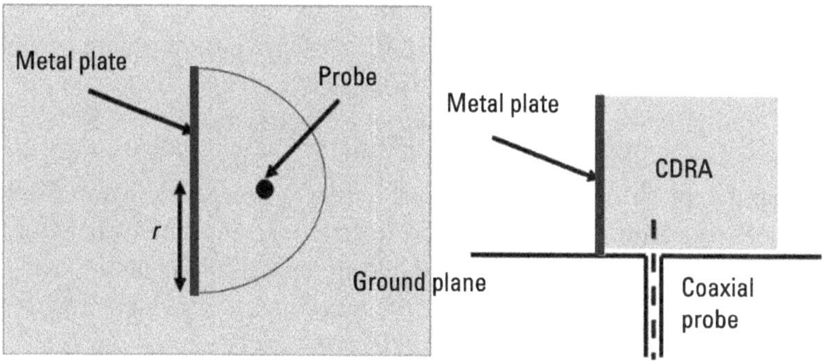

Figure 2.44 The geometry of a metal plate with half cylindrical DRA top and side view [73].

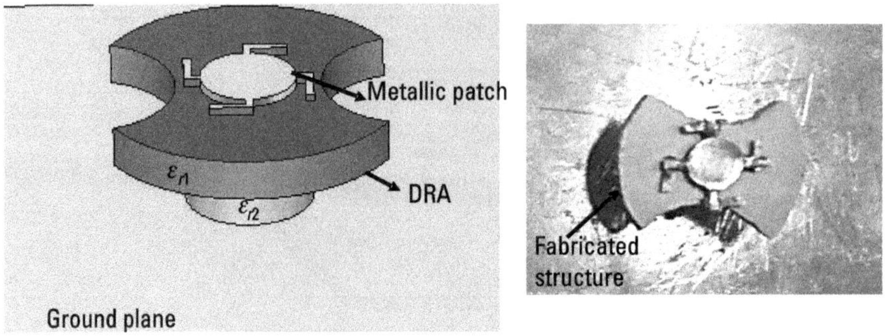

Figure 2.45 Stacked antenna geometry and fabricated structure [75].

4.4) and Rogers RO3210 (ε_r= 10.2) dielectric materials are used to fabricate the antenna.

In [76], a Koch snowflake DRA loaded with a circular metallic patch is reported for wideband applications. Applying the Koch snowflake fractal approach to the CDRA increases the surface area–to-volume ratio of the DRA, thus improving the bandwidth of the antenna. Also the circular patch behaves like a frequency tuner and shifts the frequency to lower sideband. This antenna covers a 2.61–4.72-GHz (57.5%) frequency band with a maximum 8-dBi gain.

Numerous low-profile CDRAs have been reported for wireless communication systems [77–85]. Most of the low-profile antennas are made using high dielectric material. In [77], a low-profile circular disk DR antenna is constructed with a very high dielectric permittivity material (ε_r= 82), and it offers low resonant frequency. For the TM_{101} mode, an antenna is designed and investigated. The resonant frequency is calculated by

$$f_{110}^{TM} \cong \frac{c}{4h\sqrt{\varepsilon_r}} \tag{2.27}$$

where c is the light velocity, h is the height of the low-profile antenna, and ε_r is the high dielectric constant. Figure 2.46 shows a low-profile antenna geometry.

In [78], a CDRA of high permittivity is excited by a microstrip line. The antenna configuration has the advantage of being low-profile and offers a 3.6% impedance bandwidth, which is useful for industrial, scientific, and medical (ISM) band applications. In [79], a small dielectric antenna is investigated for a new wireless communication system. By selecting the appropriate height for the metallic cap, a low-profile CDRA with a lower operating frequency can be easily designed. The CDRA is covered by the metallic cap. It reduces the height of the antenna but simultaneously reduces the resonant frequency by 30.6%. This antenna design offers a solution for making a low-profile antenna rather than

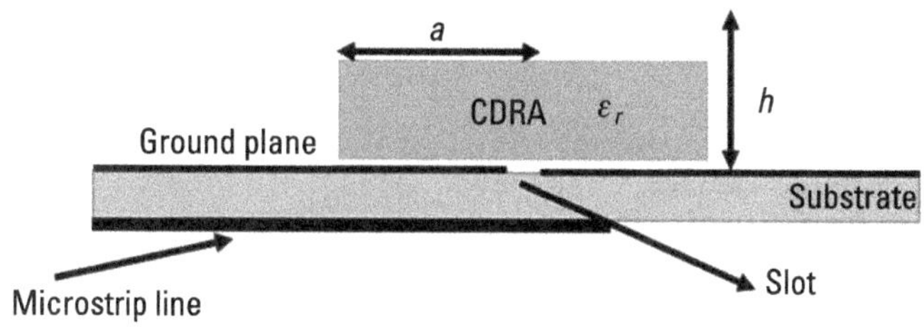

Figure 2.46 Low-profile CDRA geometry, side view.

using a high dielectric permittivity material. Figure 2.47 shows the aperture-coupled small low-profile CDRA geometry.

In [80], to design a low-profile antenna, a very low (length-to-height) aspect ratio ≈ 6 is selected. It is constructed with a $\varepsilon_r = 10.8$ high-permittivity material; the antenna is excited by aperture coupling through a microstrip feed line that is perfectly matched to the 50-Ω impedance. Figure 2.48 shows this low-profile dielectric antenna configuration.

In [81], a perforation and edge-grounding technique is employed to improve the bandwidth performance of the DRA. Metal is placed on the chopped edge of DRA, and the resonant frequency shifts to 2.82 and 3.89 GHz, respectively, and they merge to offer a wide frequency band. To create perforations, an array of symmetrical square-shaped slots is drilled into the DRA. This leads to a decreased Q factor of the antenna, which enhances the operating bandwidth range. This antenna geometry offers a 56% wide impedance bandwidth for various wireless applications.

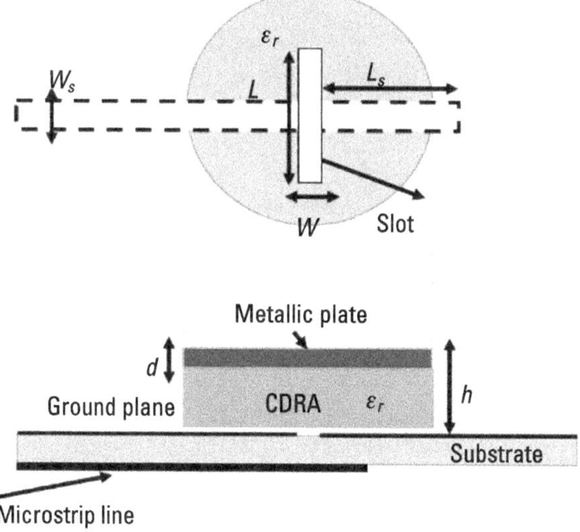

Figure 2.47 Aperture-coupled small low-profile CDRA, top and side view [79].

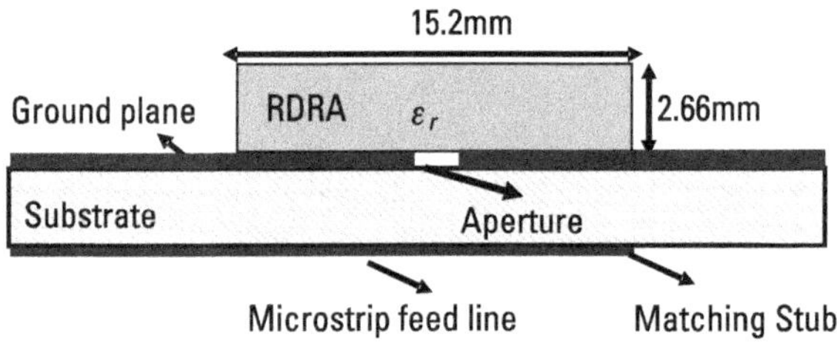

Figure 2.48 Low-profile dielectric antenna configuration [80].

In order to design a compactly sized antenna, an HDRA with complementary split rings and slots is investigated for wideband applications [82]. The complementary split-ring resonator (CSRR)-shaped slots are created on an HDRA to lower the Q factor of the antenna, thus improving the bandwidth. Subsequently, the CSRRs are etched from the ground plane, generating the current cancelation in the CSRR, thus reducing cross-polar components. A 30% bandwidth is obtained at the resonant frequency of 1.9 GHz. Figure 2.49 depicts the fabricated structure of this CSRR-loaded HDRA.

2.5 Fractal- and Reform-Shaped Geometry

This section discusses bandwidth enhancement by application of the fractal approach on the DRA. Figure 2.50 shows various modified geometries [83–92].

The antenna surface area–to-volume ratio can be increased by employing a fractal approach on the DRA or reformed geometry. Due to less base area contact between the DRA and the ground plane, maximum current coupling occurs to DRA through various feeding techniques. Antenna radiation losses increase because of increased reflection at the boundary of the dielectric-air interface due to dielectric permittivity discontinuity at the interface, thereby significantly improving impedance bandwidth [31].

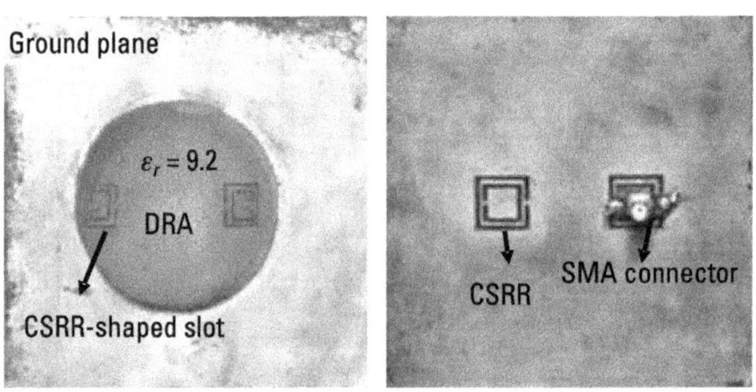

Figure 2.49 The fabricated structure of a CSRR-loaded HDRA [82].

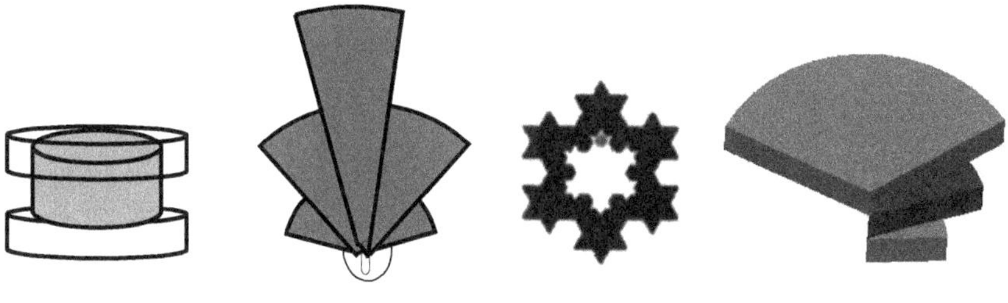

Figure 2.50 Various modified CDRA geometries [83–92].

Other important techniques for enhancing the bandwidth include reducing the surface area–to-volume ratio (*S*/*V*) of the antenna, exciting mode matching at the resonant frequency, diminishing the base area contact of the DRA to the ground plane, and reducing the effective dielectric constant value and perturbation through a decrease in the Q factor. The resonant frequency is shifted by the perturbation created on the cavity or the DRA [83]. Through (2.28), the effect on the frequency range can be verified for volume perturbation of antenna.

$$\frac{f_r - f_0}{f_r} = \frac{-\iiint\left(\Delta\varepsilon\ \mathrm{E}.E_0^* + \Delta\mu\ \mathrm{H}.H_0^*\right).\partial\tau}{\iiint\left(\varepsilon\ \mathrm{E}.E_0^* + \mu\ \mathrm{H}.H_0^*\right).\partial\tau} \tag{2.28}$$

If a small sample perturbation is created on the cavity, then E field E_0 and H field H_0 exist in the unperturbed state, and the fields in the interior of the cavity are *E* and *H*. In (2.28), f_r is resonant frequency; ε and μ are the permittivity and permeability of the unperturbed cavity, respectively; ∂_r is the elementary volume; and $\Delta\varepsilon$ and $\Delta\mu$ are the changes in the permittivity and permeability due to the insertion of perturbation in the cavity. Many other modifications of CDRA geometries increase radiation losses, which contributes to enhancing the bandwidth of antenna. Numerous fractal-based DRA and modified DRA geometries are reported for achieving wide bandwidth with compactness [84–93].

A fractal is a self-similar repetitive pattern–based geometry, which can increase the radiation boundary of the DRA for receiving and transmitting electromagnetic signals. Additionally, increasing the surface area–to-volume ratio of the antenna decreases the Q factor and improves the overall impedance bandwidth. In [84], a dual-element RDRA is investigated for reducing the size and wideband applications by combining Sierpinski and Minkowski fractals. It is made using dielectric material Rogers RT6010 (ε_{r1} = 10.2) and dielectric material FR-4 (ε_{r2} = 4.3). Applying Sierpinski and Minkowski fractals of every iteration in the DRA reduces the Q factor and enhances the surface area–to-volume ratio, which enhances the bandwidth with gain, which is calculated by

$$B.W. = \frac{1}{Q_{factor}} \propto \frac{1}{2w_0(\varepsilon_r)^p}\left(\frac{surface}{volume}\right)^{-s} \tag{2.29}$$

This antenna offers a 3.52-GHz (66%) wide impedance bandwidth with a 6.74-dBi maximum gain at a 6.25-GHz resonant frequency. Figure 2.51 describes the antenna construction process.

A fractal approach–based cylindrical segment dielectric resonator antenna is inspected for ultra-wideband applications in [85]. Stacked rotated fractal and segmented fractal half-CDRA is fabricated by FR-4 dielectric material. After each iteration of the fractal process, the surface area–to-volume (*S*/*V*) ratio increases. Thus, the Q factor decreases and the overall bandwidth improves, which is verified by the following [86].

$$B.W. = \frac{VSWR - 1}{Q\sqrt{VSWR}} \tag{2.30}$$

Here, VSWR is the voltage standing wave ratio, Q is the Q factor, and B.W. is operating bandwidth. Figure 2.52 shows stacked rotated fractal and segmented fractal half-CDRA fabricated structures. Stacked and segmented antenna geometry covers the 4.8–20-GHz (122.5%), 8.6–15.4-GHz (51.66 %) operating bandwidths with a 8.57-dBi and 15.65-dBi maximum gain, respectively, for ultra-wideband applications. Similarly in [87], fractal CDRA geometry covers a 3.6 GHz–12.8-GHz operating bandwidth range and attains a maximum gain of 9.45 dBi. It is also suitable for satellite communication and WiMAX applications.

Further, a modified fractal approach on CDRA geometry is reported in [88–90] for wideband applications. In [91], Vastu Purusha mandala (VPM)-based fractal geometry with CDRA is explored for achieving wideband antenna characteristics. This antenna geometry increases the surface area–to-volume

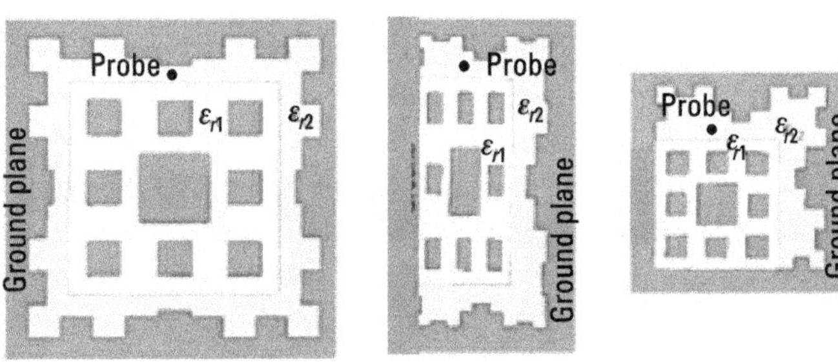

Figure 2.51 Dual-element antenna construction progress [84].

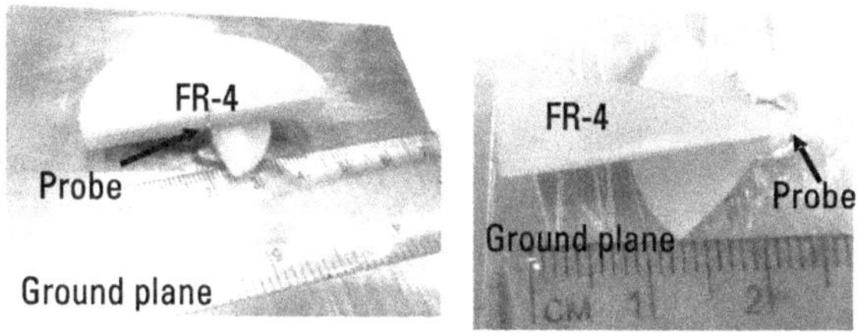

Figure 2.52 Stacked and segmented fabricated antenna structure [85].

ratio of the radiating portion of antenna. Hence, it increases the operating range of frequency for wireless applications. Figure 2.53 shows a VPM-based fractal geometry with a CDRA. Here, the antenna is made using alumina ($\varepsilon_r = 9.8$, $\tan\delta = 0.02$) dielectric material. Excitation of the antenna through an asymmetrical coaxial feed generates $TM_{10\delta}$, $TM_{20\delta}$, and $TM_{30\delta}$ radiating modes. It works on a 2.6–4.34-GHz frequency range and has a 6-dBi peak gain for WiMAX applications.

Another interesting geometry of DRA is produced by the Apollonian gasket of circles fractal approach to HDRA for improvement of the antenna characteristics [92]. The Apollonian gasket is a fractal generated from triplets of circles, where each circle is a tangent to the other two. For antenna fabrication, Rogers TMM 10 ($\varepsilon_r = 9.2$) dielectric material has been used. The surface area-to-volume ratio increases, hence reducing the Q factor to improve the bandwidth significantly. Figure 2.54 illustrates the footprint of the Apollonian gasket of circles and the fabricated structure of the antenna. The fractal geometry offers a 1.68-GHz (47%) wide impedance bandwidth with a broadside maximum gain of 8.5 dBi.

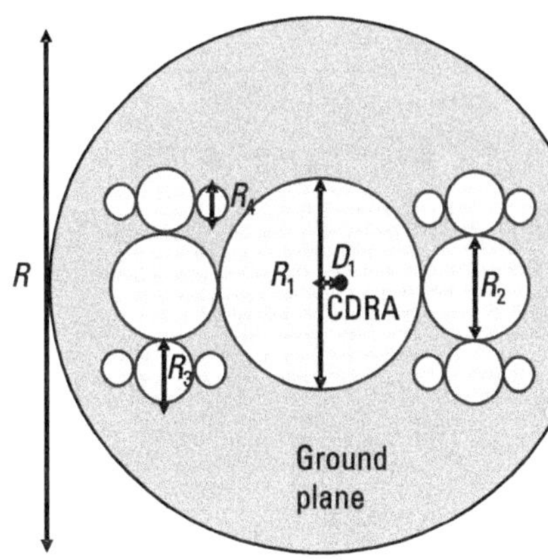

Figure 2.53 VPM-based fractal geometry with a CDRA, top view [91].

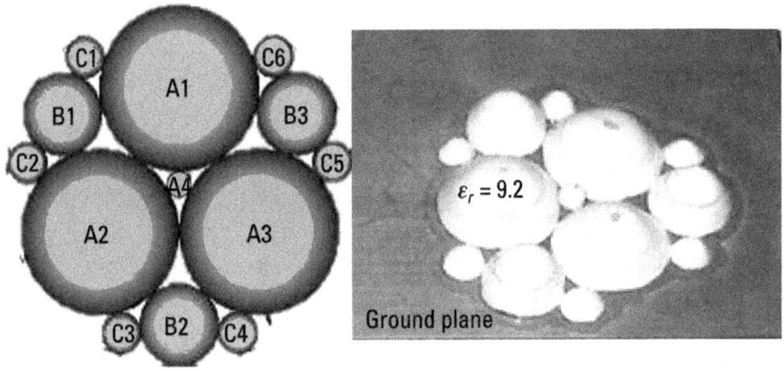

Figure 2.54 Footprint of Apollonian gasket of circles and the fabricated structure of the antenna [92].

Subsequently, the Sierpinski fractal approach is applied on an inverted pyramid–shaped DRA for wideband applications [93]. A Sierpinski fractal applied on each iteration step of the DRA reduces the base area. If the base area of the DRA reduces, more current gets coupled to the DRA, increasing the radiation losses caused due to reflection at the dielectric air interface, accounting for large permittivity discontinuity at interface and hence improving the bandwidth significantly. Also, the fractal employs a subtractive approach, which reduces the volume of the DRA with successive iterations. The fractal technique offers enhanced bandwidth by reducing the Q factor due to the increased surface area–to-volume ratio (S/V). Figure 2.55 presents the Sierpinski fractal approach–based inverted pyramidal–shaped DRA. This antenna geometry covers a 66% impedance bandwidth from 5.3 to 10.5 GHz and is suitable for C band and X band wireless applications.

Figure 2.56 shows a CDRA with a super-shaped geometry, which is designed with plastic material for wideband applications [94]. For designing a

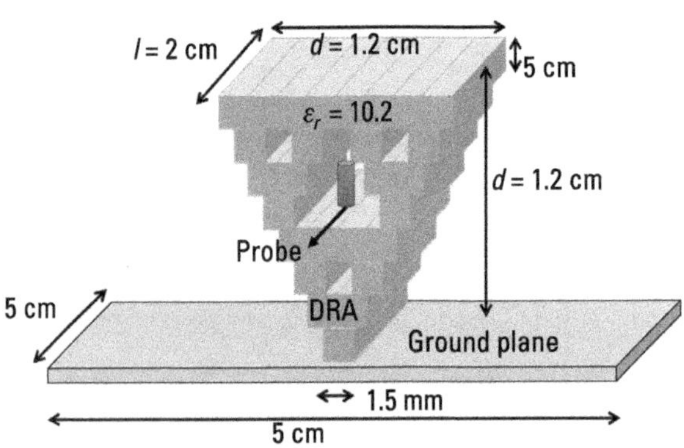

Figure. 2.55 Sierpinski fractal approach–based inverted pyramidal–shaped DRA.

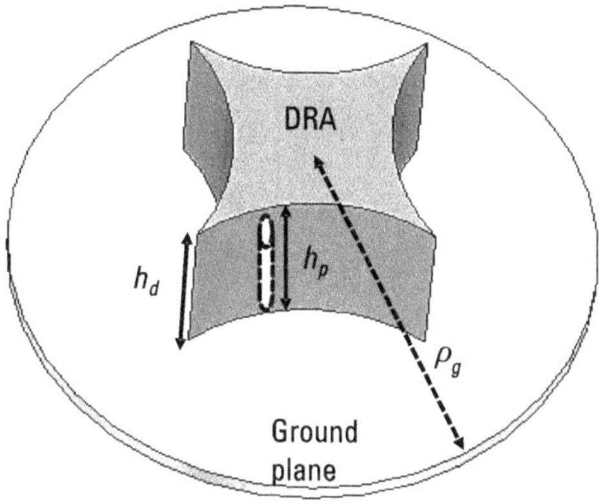

Figure 2.56 Super-shaped CDRA geometry [94].

super-shaped antenna, the machining on ceramic materials has some limitations, and the manufacturing process is time-consuming. So solid black PVC plastic is selected, with a dielectric permittivity $\varepsilon_r = 2.73$ and loss tangent tanδ = 0.33. The advantages of this material include design flexibility, outstanding electrical properties, an easy manufacturing process, and cost-effectivity. This antenna is mounted on an aluminium-based circular ground plane and excited by a coaxial probe feed. Due to the antenna geometry, a linear and circular polarized wave is generated. The fabricated prototype antenna, which has been simulated and experimentally verified, covers more than 70% impedance bandwidth and offers a stable gain and radiation pattern with high efficiency for multimedia radio systems and satellite radio receivers.

In [95], an annular-shaped CDRA is investigated for X band applications. The structure is fabricated using Rogers 03010 ($\varepsilon_r = 10.2$) dielectric material. Optimized design parameters selected to fabricate the antenna include an inner radius $r_1 = 5$ mm, an outer radius $r_2 = 10$ mm, and a height $H = 8$ mm. To modify the CDRA geometry, $\alpha = 5°$ leaned cut with the angle is incorporated on the top of the surface of the cylinder. Furthermore, a monopole of $L = 6$ mm length is inserted in the CDRA, and deformed ground plane techniques have been used. Figure 2.57 shows the geometry of this proposed antenna [95]. The insertion of a monopole and deformed ground plane in a CDRA achieves impedance matching and covers an 8.61–11.77-GHz operating bandwidth.

For achieving wideband, dumbbell-shaped [96] and corrugated plus–shaped inverted CDRA [97] have been explored. For designing a corrugated plus–shaped DRA, Rogers TMM10 ($\varepsilon_r = 10.2$) dielectric material is used and excited through a coaxial feeding network. Two corrugated strips are created on the top of a half-inverted CDRA. For impedance matching, less base area

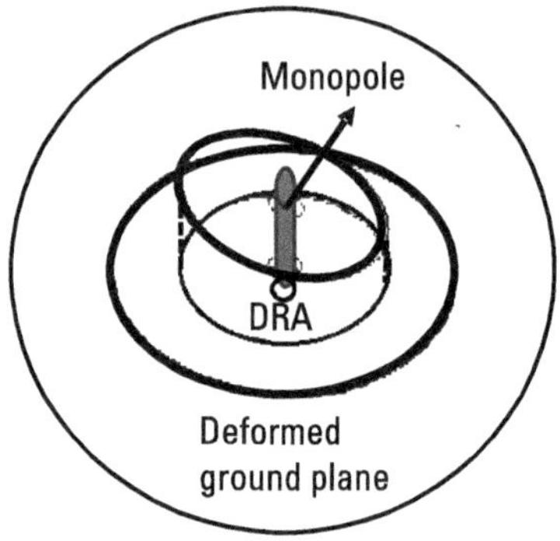

Figure 2.57 Annular-shaped antenna geometry [95].

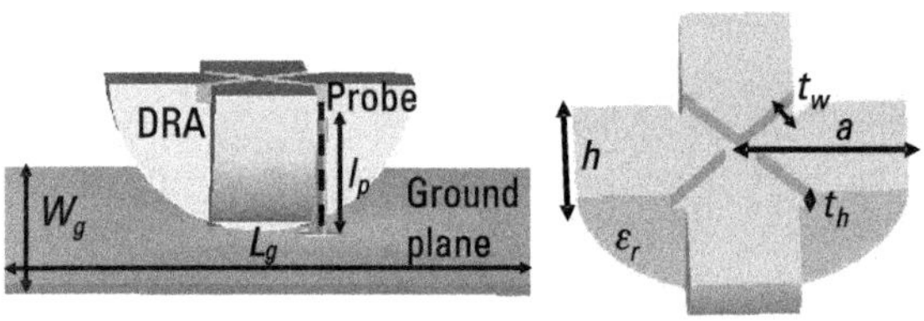

Figure. 2.58 Corrugated plus–shaped DRA geometry [94].

contact of the DRA to the ground plane and corrugated strips is needed. In this antenna $HE_{11\delta}$, $HE_{12\delta}$, HE_{123}, and HE_{142} and various hybrid modes are generated, helping to increase gain. Figure 2.58 presents the corrugated plus–shaped DRA geometry. The antenna covers 4.5-GHz (4–5.5 GHz), 6.75-GHz (6.6–6.9 GHz), and 7.9-GHz and 11.5-GHz (7.7–11.8 GHz) multiband operating frequency bands and resonant frequencies, respectively. This antenna is also suitable for IEEE 802.11p standard wireless access in vehicular environments (WAVE) applications.

2.6 Conclusion

This chapter discussed the configuration of various dielectric resonator antennas and designing techniques to achieve wide bandwidth for numerous wireless applications. Techniques detailed included various feeding techniques, the stacking of various dielectric layers, metallic loading, and compact and low-profile and fractal and reformed geometry.

References

[1] Petosa, A., *Dielectric Resonator Antenna Handbook,* Norwood, MA: Artech House, 2007.

[2] Sethares, J. C., and S. J. Naumann, "Design of Microwave Dielectric Resonators," *IEEE Transactions on Microwave Theory and Techniques,* Vol. 14, No. 1, January 1966, pp. 2–7.

[3] Mongia, R. K., and P. Bhartia, "Dielectric Resonator Antennas—A Review and General Design Relations for Resonant Frequency and Bandwidth," *International Journal of Microwave and Millimeter-Wave Computer-Aided Engineering,* Vol. 4, No. 3, 1994, pp. 230–247.

[4] Mongia, R. K., and A. Ittipiboon, "Theoretical and Experimental Investigations on Rectangular Dielectric Resonator Antennas," *IEEE Transactions on Antennas and Propagation,* Vol. 45, No. 9, Sept. 1997, pp. 1348–1356.

[5] Kishk, A.A., A.W. Glisson, and G. P. Junker, "Study of Broadband Dielectric Resonator Antennas," 1999 Antenna Applications Symposium, Alleerton Park, Monticello, IL, Sept. 1999, pp. 45–68.

[6] Gastine, M., L. Courtois, and J. L. Dormann, "Electromagnetic Resonances of Free Dielectric Spheres," *IEEE Transactions on Microwave Theory and Techniques,* Vol. 15, No. 12, December 1967, pp. 694–700.

[7] Kishk, A. A., G. Zhou, and A. W. Glisson, "Analysis of Dielectric-Resonator Antennas with Emphasis on Hemispherical Structures," *IEEE Antennas and Propagation Magazine,* Vol. 36, No. 2, April 1994, pp. 20–31.

[8] Liang, X., and T. A. Denidni, "Wideband Rectangular Dielectric Resonator Antenna with a Concave Ground Plane," *IEEE Antennas and Wireless Propagation Letters,* Vol. 8, 2009, pp. 367–370.

[9] Denidni, T. A., and Z. Weng, "Rectangular Dielectric Resonator Antenna for Ultra-wideband Applications," *IET Electronics Letters,* Vol. 45, No. 24, November 19, 2009, pp. 1210–1212.

[10] De Young, C. S., and S. A. Long, "Wideband Cylindrical and Rectangular Dielectric Resonator Antennas," *IEEE Antennas and Wireless Propagation Letters,* Vol. 5, 2006, pp. 426–429.

[11] Chu, L. C. Y., D. Guha, and Y. M. M. Antar, "Conformal Strip-Fed Shaped Cylindrical Dielectric Resonator: Improved Design of a Wideband Wireless Antenna," *IEEE Antennas and Wireless Propagation Letters,* Vol. 8, 2009, pp. 482–485.

[12] Dileep, S., V. Kiran, and B. Mukherjee, "Laterally Placed CDRA with Triangular Notches for Ultra-Wideband Applications," *Frequenz-Journal of RF Engineering and Telecommunications,* Vol. 72, Issue 1–2, 2018, pp. 73–78.

[13] Ranjan, P., and R. K. Gangwar, "Two Element Half Split Cylindrical Dielectric Resonator Antenna for Wideband Monopole-Like Radiation," *2013 IEEE Applied Electromagnetics Conference (AEMC),* Bhubaneswar, 2013, pp. 1–2.

[14] Guha, D., and Y. M. M. Antar, "Four-Element Cylindrical Dielectric Resonator Antenna for Wideband Monopole-Like Radiation," *IEEE Transactions on Antennas and Propagation,* Vol. 54, No. 92, Sept. 2006, pp. 2657–266.

[15] Singh, A., and S. K. Sharma, "Investigations on Wideband Cylindrical Dielectric Resonator Antenna with Directive Radiation Patterns and Low Cross Polarization," *IEEE Transactions on Antennas and Propagation*, Vol. 58, No. 5, May 2010, pp. 1779–1783.

[16] Mukherjee, B., P. Patel, J. Mukherjee, "A Novel Cup-shaped Inverted Hemispherical Dielectric Resonator Antenna for Wideband Applications," *IEEE Antenna and Wireless Propagation Letters*, Vol. 12, 2013, pp. 1240–1243.

[17] Kishk, A. A., Y. Yin, and A. W. Glisson, "Conical Dielectric Resonator Antennas for Wideband Applications," *IEEE Trans. Antennas Propag.*, Vol. 50, No. 4, Apr. 2002, pp. 469–474.

[18] Kishk, A., "Wideband Truncated Tetrahedron Dielectric Resonator Antenna Excited by a Coaxial Probe," *IEEE Trans. Antennas Propag.*, Vol. 51, No. 10, Oct. 2003, pp. 2913–2917.

[19] Kranenburg, R. A., and S. A. Long, "Microstrip Transmission Line Excitation of Dielectric Resonator Antennas," *IET Electronics Letters*, Vol. 24, No. 18, 1 Sept. 1988, pp. 1156–1157.

[20] Li, B., and K. W. Leung, "On the Differentially Fed Rectangular Dielectric Resonator Antenna," *IEEE Transactions on Antennas and Propagation*, Vol. 56, No. 2, Feb. 2008, pp. 353–359.

[21] Rashidian, A., and D. M. Klymyshyn, "On the Two Segmented and High Aspect Ratio Rectangular Dielectric Resonator Antennas for Bandwidth Enhancement and Miniaturization," *IEEE Transactions on Antennas and Propagation,* Vol. 57, No. 9, Sept. 2009, pp. 2775–2780.

[22] Kshirsagar, P., S. Gupta, and B. Mukherjee, "Novel Design of Conformal-Strip Excited Asymmetrical Rectangular Dielectric Resonator Antenna for Ultra-wide Band Application," *Journal of Microwave Power and Electromagnetic Energy, Taylor & Francis,* Vol. 52, No. 2, 2018, pp. 128–141.

[23] Drossos, G., Z. Wu, and L. E. Davis, "The Air Gap Effect on a Microstrip-Coupled Cylindrical Dielectric Resonator Antenna," *Microwave and Optical Technology Letters,* Vol. 20, No. 1, 1999, pp. 36–40.

[24] Kumar, A. V. P., et al., "Microstripline-Fed Halfcylindrical Dielectric Resonator Antenna For 2.4-Ghz Wlan Application," *Microwave and Optical Technology Letters,* Vol. 48, No. 4, April 2006.

[25] Majeed, A. H., et al., "Balanced Dual-Segment Cylindrical Dielectric Resonator Antennas for Ultra-Wideband Applications," *IET Microwaves, Antennas & Propagation,* Vol. 9, No. 13, pp. 1478–1486.

[26] Praveen Kumar, A. V., et al., "Microstripline Fed Cylindrical Dielectric Resonator Antenna with a Coplanar Parasitic Strip," *Progress In Electromagnetics Research,* Vol. 60, 2006, pp.143–152.

[27] Prachi, R. K. C., and K. V. Srivastava, "Wideband Cylindrical Dielectric Resonator Antenna Excited by a Rounded Bevel Shaped Patch," *2013 IEEE Antennas and Propagation Society International Symposium (APSURSI),* Orlando, FL, 2013, pp. 1910–1911.

[28] Sharma, A., and R. K. Gangwar, "Compact Quad-Band Cylindrical Dielectric Resonator Antenna with Complementary C-Shaped Defected Ground Structure," *Microwave and Optical Technology Letters,* Vol. 58, No. 3, 2016, pp. 611–615.

[29] Sharma, A., et al., "Novel Feeding Mechanism to Stimulate Triple Radiating Modes in Cylindrical Dielectric Resonator Antenna," *IEEE Access,* Vol. 4, 2016, pp. 9987–9992.

[30] Leung, K. W., et al., "Annular Slot-Coupled Dielectric Resonator Antenna," *IET Electronics Letters,* Vol. 34, No. 13, June 25, 1998, pp. 1275–1277.

[31] Huang, C.-Y., and C.-F. Yang, "Cross-Aperture Coupled Circularly Polarized Dielectric Resonator Antenna," *IEEE Antennas and Propagation Society International Symposium, 1999 Digest,* Held in conjunction with: USNC/URSI National Radio Science Meeting (Cat. No.99CH37010), Orlando, FL, 1999, Vol. 1, pp. 34–37.

[32] Huang, C.-Y., J.-Y. Wu, and K.-L. Wong, "Cross-Slot-Coupled Microstrip Antenna and Dielectric Resonator Antenna for Circular Polarization," *IEEE Transactions on Antennas and Propagation,* Vol. 47, No. 4, April 1999, pp. 605–609.

[33] Luk, K. M., and K. W. Leung, *Dielectric Resonator Antennas,* Baldock, England: Research Studies Press Ltd., 2003.

[34] Kishk, A. A., et al., "Slot excitation of the Dielectric Disk Radiator," *IEEE Transactions on Antennas and Propagation,* Vol. 43, No. 2, Feb. 1995, pp. 198–201.

[35] Buerkle, A., K. Sarabandi, and H. Mosallaei, "Compact Slot and Dielectric Resonator Antenna with Dual-Resonance, Broadband Characteristics," *IEEE Transactions on Antennas and Propagation,* Vol. 53, No. 3, March 2005, pp. 1020–1027.

[36] Kumar, R., and R. K. Chaudhary, "A Dual-Band Dual-Polarized Cubical DRA Coupled with New Modified Cross-shaped Slot for ISM (2.4 GHz) and Wi-MAX (3.3–3.6 GHz) Band Applications," *International Journal of RF and Microwave Computer-Aided Engineering (RFMiCAE),* Wiley Journals, Vol. 29 (1), e21449, 2018.

[37] Drossos, G., Z. Wu, and L. E. Davis, "Aperture-Coupled Cylindrical Dielectric Resonator Antenna," *Microwave and Optical Technology Letters,* Vol. 20, Issue 6, 1999, pp. 407–414.

[38] Huang, T.-H., et al., "A Simple Design Cylindrical Dielectric Resonator Antenna with Wide Aperture-Coupled for Broadband Applications," *Microwave and Optical Technology Letters,* Vol. 49, Issue 5, pp. 1064–1067.

[39] Majeed, A. H., et al., "Aperture-Coupled Asymmetric Dielectric Resonators Antenna for Wideband Applications," *IEEE Antennas and Wireless Propagation Letters,* Vol. 13, 2014, pp. 927–930.

[40] Zebiri, C., et al., "Offset Aperture-Coupled Double-Cylinder Dielectric Resonator Antenna with Extended Wideband," *IEEE Transactions on Antennas and Propagation,* Vol. 65, No. 10, Oct. 2017, pp. 5617–5622.

[41] Sun, Y. X., and K. W. Leung, "Dual-Band and Wideband Dual-Polarized Cylindrical Dielectric Resonator Antennas," *IEEE Antennas and Wireless Propagation Letters,* Vol. 12, 2013, pp. 384–387.

[42] Sharma A., G. Das, and R. K. Gangwar, "Dual-Band Dual-Polarized Hybrid Aperture Cylindrical Dielectric Resonator Antenna for Wireless Applications," *Int J RF MicrowComput Aided Eng.,* 2017.

[43] Leung, K.-W., K.-M. Luk, and K. Y. A. Lai, "Theory and Experiment of an Aperture-Coupled Hemispherical Dielectric Resonator Antenna," *IEEE Trans Antennas Propag.*, Vol. 43, No. 11, 1995, pp. 1192– 1198.

[44] Kishk, A. A., B. Ahn and D. Kajfez, "Broadband Stacked Dielectric Resonator Antennas," *IET Electronics Letters*, Vol. 25, No. 18, 1989, pp. 1232–1233.

[45] Fakhte, S., H. Oraizi, and R. Karimian, "A Novel Low-Cost Circularly Polarized Rotated Stacked Dielectric Resonator Antenna," *IEEE Antennas and Wireless Propagation Letters*, Vol. 13, 2014, pp. 722–725.

[46] Bit-Babik, G., C. Di Nallo, and A. Faraone, "Multimode Dielectric Resonator Antenna of Very High Permittivity," *IEEE Antennas and Propagation Society Symposium,* Monterey, CA, Vol. 2, 2004, pp. 1383–1386.

[47] Chauhan, M., and B. Mukherjee, "Investigation of T-Shaped Compact Dielectric Resonator Antenna for Wideband Application," *Radio Electron Commun Syst.,* Vol. 62, No. 11, 2019, pp. 594–603.

[48] Rashidian, A., K. Forooraghi, and M. T. Aligodarz, "Investigations of Two Segment Dielectric Resonator Antennas," *Microwave Opt Technol Lett.*, Vol. 45, No. 6, June 2005, pp. 533–537.

[49] Kshirsagar, P., S. Gupta, and B. Mukherjee, "A Two Segment Rectangular Dielectric Resonator Antenna for Ultra-Wideband Application," *Electromagnetics, Taylor & Francis Electromagnetics, Taylor & Francis,* Vol. 38, Issue 1, pp. 20–33, 2018.

[50] Walsh, A. G., C. S. DeYoung, and S. A. Long, "An Investigation of Stacked and Embedded Cylindrical Dielectric Resonator Antennas," *IEEE Antennas and Wireless Propagation Letters,* Vol. 5, 2006, pp. 130–133.

[51] Ranjan, P., and R. K. Gangwar, "Segmented Quarter-Cylindrical Dielectric Resonator Antenna: Simulation and Experimental Investigation In Composite Form for Wideband Applications," *International Journal of Microwave and Wireless Technologies,* Vol. 9, No. 4, 2016, pp. 881–890.

[52] Sharma, A., and R. K. Gangwar, "Asymmetrical Annular Shape Microstrip Line Fed Stacked Cylindrical Dielectric Resonator Antenna for UWB," *2016 Asia-Pacific Microwave Conference (APMC),* New Delhi, India, 2016, pp. 1–4.

[53] Chaudhary, R. K., K. V. Srivastava, and A. Biswas, "Two-Layer Embedded Half-Split Cylindrical Dielectric Resonator Antenna for Wideband Applications," *2012 42nd European Microwave Conference, Amsterdam,* 2012, pp. 815–817.

[54] Chaudhary, R. K., K. V. Srivastava, and A. Biswas, "Broadband Four-Element Multilayer Multipermittivity Cylindrical Dielectric Resonator Antenna," *Microwave and Optical Technology Letters,* Vol. 55, No. 4, 2013, pp. 932–937.

[55] Chaudhary, R. K., et al., "Coaxial Fed Half-Split Multilayer Cylindrical Dielectric Resonator Antenna for Wideband Applications," *2011 41st European Microwave Conference,* Manchester, England, 2011, pp. 1015–1018.

[56] Chaudhary, R. K., K. V. Srivastava, and A. Biswas, "A Concentric Three-Layer Half-Split Cylindrical Dielectric Resonator Antenna for Wideband Applications," *2013 International Symposium on Electromagnetic Theory,* Hiroshima, Japan, 2013, pp. 664–667.

[57] Chaudhary, R. K., K. V. Srivastava, and A. Biswas, "Variation of Permittivity in Radial Direction in Concentric Half-Split Cylindrical Dielectric Resonator Antenna for Wideband Application," *International Journal of RF and Microwave Computer-Aided Engineering,* Vol. 25, No. 4, 2014, pp. 321–329.

[58] Chaudhary, R. K., K. V. Srivastava, and A. Biswas, "A Practical Approach: Design of Wideband Cylindrical Dielectric Resonator Antenna with Permittivity Variation in Axial Direction and its Fabrication Using Microwave Laminates," *Microwave and Optical Technology Letters,* Vol. 55, No. 10, 2013, pp. 2282–2288.

[59] Fang, X. S., and K. W. Leung, "Design of Wideband Omnidirectional Two-Layer Transparent Hemispherical Dielectric Resonator Antenna," *IEEE Trans Antennas Propag.*, Vol. 62, No. 10, 2014, pp. 5353–5357.

[60] Guha, D., et al., "Segmented Hemispherical DRA: New Geometry Characterized and Investigated in Multi-Element Composite Forms for Wideband Antenna Applications," *IEEE Transactions on Antennas and Propagation,* Vol. 60, No. 3, March 2012, pp. 1605–1610.

[61] Lapierre, M., et al., "A Wideband Monopole Antenna Using Dielectric Resonator Loading," *IEEE Antennas and Propagation Society International Symposium. Digest. Held in conjunction with: USNC/CNC/URSI North American Radio Sci. Meeting* (Cat. No.03CH37450), Columbus, OH, 2003, pp. 16–19.

[62] Guha, D., et al., "Improved Design Guidelines for the Ultra Wideband Monopole-Dielectric Resonator Antenna," *IEEE Antennas and Wireless Propagation Letters,* Vol. 5, 2006, pp. 373–376.

[63] Bijumon, P. V., et al., "Enhanced Bandwidth Microstrip Patch Antennas Loaded with High Permittivity Dielectric Resonators," *Microwave and Optical Technology Letters,* Vol. 35, No. 4, 2002, pp. 327–330.

[64] George, J., et al., "Dielectric-Resonator-Loaded Microstrip Antenna for Enhanced Impedance Bandwidth and Efficiency," *Microwave and Optical Technology Letters,* Vol. 17, No. 3, 2002, pp. 205–207.

[65] Lapierre, M., et al., "Ultra Wideband Monopole/Dielectric Resonator Antenna," *IEEE Microwave and Wireless Components Letters,* Vol. 15, No. 1, Jan. 2005, pp. 7–9.

[66] Ozzaim, C., "Monopole Antenna Loaded by a Stepped-Radius Dielectric Ring Resonator for Ultrawide Bandwidth," *IEEE Antennas and Wireless Propagation Letters,* Vol. 10, 2011, pp. 843–845.

[67] Yung, E. K. N., W. W. S. Lee, and K. M. Luk, "Microstrip Antenna Top-Loaded by a Dielectric Resonator," *Microwave and Optical Technology Letters,* Vol. 7, No. 2, 1994, pp. 55–57.

[68] Esselle, K. P., "A Dielectric-Resonator-on-Patch (DROP) Antenna for Broadband Wireless Applications: Concept and Results," *Proc. IEEE Antennas and Propagation Society (AP-S) Int. Symp.*, Vol. 2, July 2001, pp. 22–25.

[69] Mukherjee, B., A Novel Half Hemispherical Dielectric Resonator Antenna with Array of Slots Loaded with a Circular Metallic Patch for Wireless Applications," *AEU–International Journal of Electronics and Communication,* Vol. 69, 2015, pp. 1755–1759.

[70] Guha, D., B. Gupta, and Y. M. M. Antar, "Hybrid Monopole-DRAs Using Hemispherical/ Conical-Shaped Dielectric Ring Resonators: Improved Ultrawide Band Designs," *IEEE Trans Antennas Propag.*, Vol. 60, No. 1, 2012, pp. 393–398.

[71] Hosseinbeig. A., M. Kamyab, and J. S. Meiguni, "Theory of Aperture-Coupled Hemispherical Dielectric Resonator Antennas with Radiating Elements," *AEU–Int J Electronics Commun,* Vol. 67, No. 11, 2013, pp. 975–980.

[72] Ge, Y., K. P. Esselle, and T. S. Bird, "Compact Dielectric Resonator Antennas with Ultrawide 60%–110% Bandwidth," *IEEE Transactions on Antennas and Propagation,* Vol. 59, No. 9, Sept. 2011, pp. 3445–3448.

[73] Tam, M. T. K., and R. D. Murch, "Half Volume Dielectric Resonator Antenna Designs," *IET Electronics Letters,* Vol. 33, No. 23, Nov. 6, 1997, pp. 1914–1916.

[74] Tam, M. T. K, and R. D. Murch, "Compact Circular Sector and Annular Sector Dielectric Resonator Antennas," *IEEE Transactions on Antennas and Propagation,* Vol. 47 May 1999, pp. 837–842.

[75] Chauhan, M., A. K. Pandey, and B. Mukherjee, "A Novel Compact Cylindrical Dielectric Resonator Antenna for Wireless Sensor Network Application," *IEEE Sensors Letters,* Vol. 2, No. 2, June 2018, pp. 1–4.

[76] Narayanan, D. S., D. V. Kiran, and B. Mukherjee, "Koch Snowflake Dielectric Resonator Antenna Loaded with a Circular Metallic Patch for Wideband Applications," *2016 URSI Asia-Pacific Radio Science Conference (URSI AP-RASC),* Seoul, South Korea, 2016, pp. 1798–1801.

[77] Leung, K. W., et al., "Characteristics of a Low-Profile Circular Disk DR Antenna with Very High Permittivity," *IET Electronics Letters,* Vol. 31, No. 6, 1995, pp. 417–418.

[78] Leung, K. W., et al., "Low-Profile Circular Disk DR Antenna of Very High Permittivity Excited by a Microstripline," *IET Electronics Letters,* Vol. 33 No. 12, 1997.

[79] Lee, M. T., et al., "A Small Dielectric Resonator Antenna," *IEEE Transactions on Antennas and Propagation,* Vol. 50, No. 10, 2002, pp. 1485–1487.

[80] Esselle, K. P., "A Low-Profile Rectangular Dielectric-Resonator Antenna," *IEEE Transactions on Antennas and Propagation,* Vol. 44, No. 9, Sept. 1996, pp. 1296–1297.

[81] Patel, P., B. Mukherjee, and J. Mukherjee, "A Compact Wideband Rectangular Dielectric Resonator Antenna Using Perforations and Edge Grounding," *IEEE Antenna and Wireless Propagation Letters,* Vol. 14, 2015, pp. 490–493.

[82] Mukherjee, B., P. Patel, and J. Mukherjee, "A Novel Hemispherical Dielectric Resonator Antenna with Complimentary Split Rings Shaped Slots and Resonator for Wideband and Low Cross-Polar Applications," *IEEE Antenna and Propagation Magazine,* Vol. 57, Issue 1, 2015, pp. 120–128.

[83] Harington, R. F., *Time Harmonic Electromagnetic Fields,* New York: IEEE Press, 2001.

[84] Kiran, V., S. Dileep, and B. Mukherjee, "Compact Embedded Dual-Element Rectangular Dielectric Resonator Antenna Combining Sierpinski and Minkowski Fractals," *IEEE Transactions on Components, Packaging and Manufacturing Technology,* Vol. 7, No. 5, 2017, pp. 786–791.

[85] Gupta, S., P. Kshirsagar and B. Mukherjee, "A Low-Profile Multilayer Cylindrical Segment Fractal Dielectric Resonator Antenna: Usage for Wideband Applications," *IEEE Antennas and Propagation Magazine,* Vol. 61, No. 4, Aug. 2019, pp. 55–63.

[86] Kajfez, D., A. W. Glisson, and J. James, "Computed Modal Field Distribution for Isolated Dielectric Resonators," *IEEE Trans. Microw. Theory Techn.*, Vol. 32, Dec. 1984, pp. 1609–1616.

[87] Chauhan, M., and B. Mukherjee, "High Gain Fractal Cylindrical Dielectric Resonator Antenna for UWB Application," *2018 IEEE Radio and Antenna Days of the Indian Ocean (RADIO),* Grand Port, 2018, pp. 1–2.

[88] Sankaranarayanan, D., D. Venkatakiran, and B. Mukherjee, "A Novel Compact Fractal Ring Based Cylindrical Dielectric Resonator Antenna for Ultra Wideband Application," *Progress in Electromagnetics Research C,* Vol. 67, 2016, pp. 71–83.

[89] Sankaranarayanan, D., D. Venkatakiran, and B. Mukherjee, "Compact Bi-Cone Dielectric Resonator Antenna for Ultra-Wideband Applications—A Novel Geometry Explored," *Electromagnetics,* Vol. 37, Issue 7, 2017, pp. 471–481.

[90] Dileep S., D. V. Kiran, and B. Mukherjee, "Wideband Koch snowflake Fractal Dielectric Resonator Antenna," *2015 IEEE 10th International Conference on Industrial and Information Systems (ICIIS),* Peradeniya, 2015, pp. 205–209.

[91] Gangwar, K., et al., "Investigation on Novel Wideband Fractal Antenna Design Based on Cylindrical Shape Dielectric Resonator," *International Journal of RF and Microwave Computer-Aided Engineering,* 2019.

[92] Mukherjee, B., P. Patel, and J. Mukherjee, "Hemispherical Dielectric Resonator Antenna based on Apollonian Gasket of Circles—A Fractal Approach," *IEEE Transactions on Antennas and Propagation,* Vol. 62, No. 1, 2014, pp. 40–47.

[93] Gupta, S., P. Kshirsagar, and B. Mukherjee, "Sierpinski Fractal Inspired Inverted Pyramidal DRA for Wideband Applications," *Electromagnetics,* Vol. 38, No. 2, 2018, pp. 103–112.

[94] Simeoni, M., et al. "Plastic-Based Supershaped Dielectric Resonator Antennas for Wide-Band Applications," *IEEE Transactions on Antennas and Propagation,* Vol. 59, No. 12, Dec. 2011, pp. 4820–4825.

[95] Zhao, N., et al., "A Wideband Annular Cylindrical Dielectric Resonator Antenna for X-Band Applications," *2016 Progress in Electromagnetic Research Symposium (PIERS),* Shanghai, 2016, pp. 4005–4008.

[96] Chaudhary, R. K., K. V. Srivastava, and A. Biswas, "A Broadband Dumbbell-Shaped Dielectric Resonator Antenna," *Microwave and Optical Technology Letters,* Vol. 56, No. 12, 2014, pp. 2944–2947.

[97] Chauhan, M., A. Rajput, and B. Mukherjee, "Multiband Corrugated Plus Shaped Inverted half Cylindrical Dielectric Resonator Antenna for Numerous Wireless Applications," *2019 13th European Conference on Antennas and Propagation (EuCAP),* Krakow, Poland, 2019, pp. 1–5.

3

DRA CP Techniques

Antenna polarization is one of the key elements of antenna radiation characteristics for modern wireless communication systems. Any antenna can transmit or receive linear, circular, and elliptically polarized signals. Satellite communication, radar navigation, space communication, and telemetry tracking and command require dual or circular polarized signals. Circular polarized signals are less affected by the environment and offer orientation independence, as illustrated in Figure 3.1.

Antenna polarization is usually determined by the radiated electric field orientation. Linearly polarized antennas can radiate in a vertical or horizontal plane in the direction of the wave propagation, whereas circular polarized antennas radiate electromagnetic signals in a circular pattern that can cover horizontal and vertical planes. Due to misalignment, phasing issues, and line of sight problems between the transmitter and receiver sides, circularly polarized antennas are preferred over linearly polarized antennas for such applications. In addition, circularly polarized antennas reduce multipath problems such as reflections through features such as obstacles, buildings, and the ground. Another important advantage of circular polarized antennas is that they offer multipath rejection, which leads to signal interference in linear polarization.

- *Polarization:* The polarization of electromagnetic signals refers to the specific motions of components of the electromagnetic fields (E or H) with respect to time. If the signal propagates in the Z direction and electric field components are present in the X and Y directions, then the total time-varying propagating electric field is defined as [1]

$$E_{total} = E_x \cos(\omega t + \Phi_x)\hat{x} + E_y \cos(\omega t + \Phi_y)\hat{y} \tag{3.1}$$

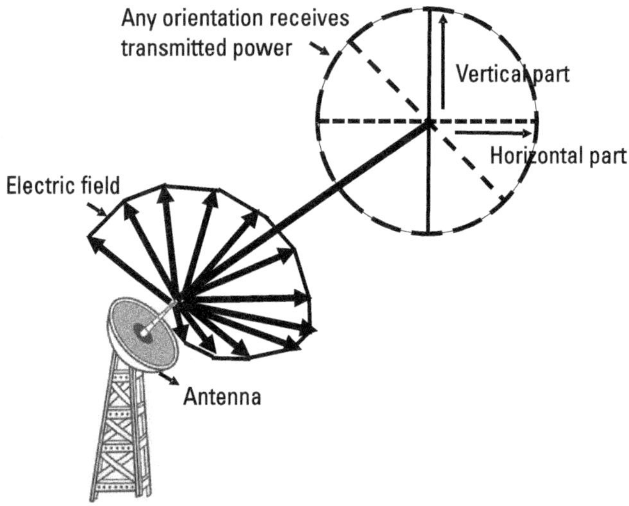

Figure 3.1 Circular polarized transmitted and received signals.

where Φ_x and Φ_y are the phase angle of electric field components of the X and Y directions, and E_x and E_y are the amplitude of electric field components of the X and Y directions. Figure 3.2 illustrates linear, circular, and elliptical polarized electromagnetic waves in free space.

- *Linear polarization:* An electromagnetic wave is linearly polarized if the EM wave field components (E_x and E_y) are oriented along the same plane (either vertical or horizontal) at a specific point.

  For linear polarization, the EM field components and phase angle accomplish the following conditions:

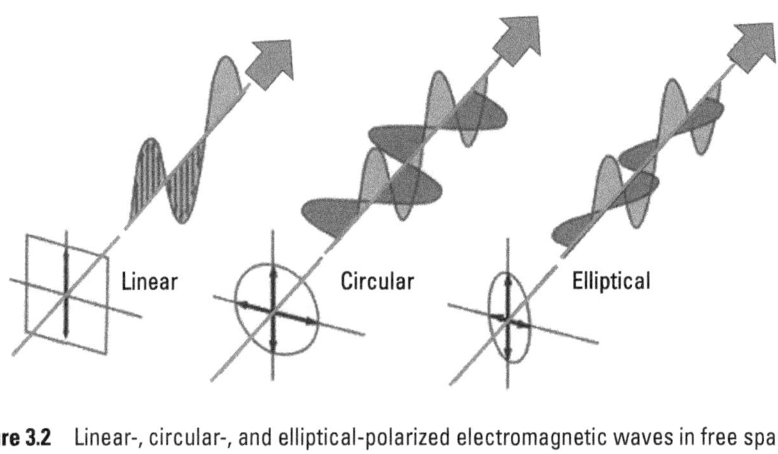

Figure 3.2 Linear-, circular-, and elliptical-polarized electromagnetic waves in free space.

1. Two orthogonal field components of the same amplitude ($E_x = E_y = E$).
2. Two orthogonal field components in the same phase ($\Phi_x = \Phi_y = \Phi_0$).

- *Circular polarization:* An EM wave is circular-polarized if the EM wave field components (E_x and E_y) are oriented along both planes (vertical and horizontal) at a specific point.

  For circular polarization, the EM field components and phase angle satisfy the following conditions:

  1. Two orthogonal field components of the same amplitude ($E_x = E_y = E$);
  2. Two orthogonal field components with a time phase difference of $\pm \pi/2 (\Phi_x - \Phi_y = \pm \pi/2)$.

  If the field components' rotation is clockwise, then the EM wave is right-hand (or clockwise) circularly polarized (RHCP). If the rotation is counterclockwise, then the EM wave is left-hand (or counterclockwise) circularly polarized (LHCP).

- *Elliptical polarization:* An EM wave is elliptical-polarized if the wave field components (E_x and E_y) are oriented along both the planes (vertical and horizontal) at a specific point. They satisfy the following conditions:

  1. Two orthogonal field components with different amplitudes ($E_x \neq E_y$);
  2. Two orthogonal field components with phase difference ($\Phi_x - \Phi_y \neq 0$).

- *Axial Ratio:* The ratio of the major axis to the minor axis of the E field component is referred to as the axial ratio (AR).

$$\mathit{Axial\ ratio}(AR)\,\mathit{in}\,dB = \frac{\mathit{major\ axis\ electric\ field\ componets}}{\mathit{minor\ axis\ electric\ field\ componets}} = 20\log\left(\frac{E_{max}}{E_{min}}\right) \quad (3.2)$$

E_{max} and E_{min} are magnitudes of the maximum and minimum of the electric field components.

The AR of the circular polarized wave is calculated from (3.3) [1].

$$\mathit{Axial\ ratio}(AR)\,\mathrm{in}\,dB = 10\log\left[\frac{E_x^2 + E_y^2 + \left(E_x^4 + E_y^4 + 2E_x^2E_y^2\cos\Delta\Phi\right)^{1/2}}{E_x^2 + E_y^2 - \left(E_x^4 + E_y^4 + 2E_x^2E_y^2\cos\Delta\Phi\right)^{1/2}}\right] \quad (3.3)$$

if $\Delta\Phi = 2(\Phi_x - \Phi_y)$ is the phase difference of the field components. If the amplitude of the field components is the same ($E_x = E_y = E$) and $\Phi_x - \Phi_y = \pm\pi/2$ (i.e., AR= 0 dB). Even so, in practical applications, AR up to –3 dB is admissible for circular-polarized applications.

Over the last few decades, many researchers have done significant work for achieving circular polarization with the DRA. The circular-polarized DRA always fascinated researchers because of its powerful key features, including more agility between the transmitter and receiver orientation angle, much fewer multipath reflections, and enhanced mobility compared to linear polarization. Several techniques have been extensively investigated in the DRA for obtaining circular polarization; they include single-feed, dual-feed, and multiple-feed excitations, various modified antenna geometries, and various antenna array configuration arrangements.

3.1 Circular Polarization by the Single-Feed Technique

This section explores various single-feed techniques, including coaxial, microstrip, and aperture with a modified ground plane [2–30]. A single-feed technique achieves circular polarization in the DRA [2], because two degenerate orthogonal modes with a quadrature phase difference are generated between the modes. Figure 3.3 plots the phase and axial ratio graphs for DRA quasi-degenerate modes. The AR of an antenna can be calculated if the amplitude value of the resonant frequency and phase difference response of degenerate modes are obtained.

In [3–20], design techniques for achieving circular polarization in a single-feed excitation DRA are explored. A single-slot fed-through RDRA for circular polarization is reported in [3, 4]. An optimized feed position through the DRA generates orthogonal modes and achieves a wide AR (1.8%) bandwidth and a wide beamwidth (110°) [3]. This RDRA is constructed using $\varepsilon_r = 40$ dielectric material and excited by the aperture microstrip feeding method. Figure 3.4(a) illustrates a slot-coupled RDRA; a 45-degree-rotated RDRA with respect to the slot in the ground plane is fixed, on a copper ground plane, and on the bottom side of a substrate, a microstrip feed line is incorporated. For the excitation of DRA, a nonresonant/nonradiating slot is etched in the ground plane for the coupling of the energy from the microstrip feed line to the DRA. For good impedance matching and AR bandwidth, slot dimensions, width (W), and length (l), are optimized. Two resonant modes TE^x_{111} and TE^y_{111} are excited in the DRA, while radiated in the X and Y directions. As well, to enhance the circular polarization in the RDRA, a (length) and b (width) dimensions of geometry satisfy (3.4) as

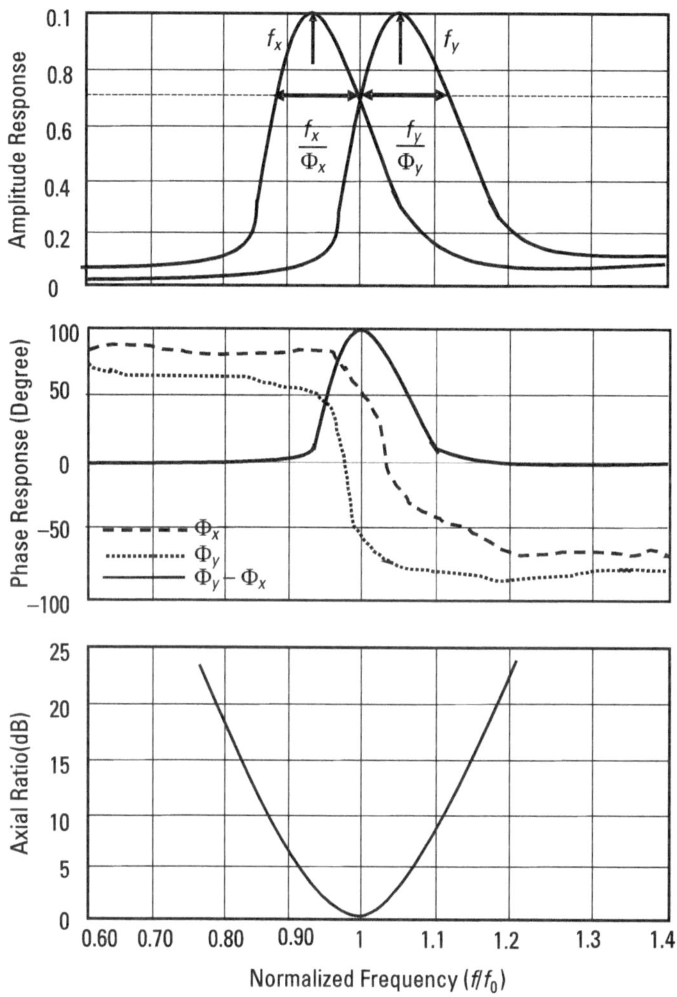

Figure 3.3 DRA quasi-degenerate modes: magnitude, phase, and axial ratio plots.

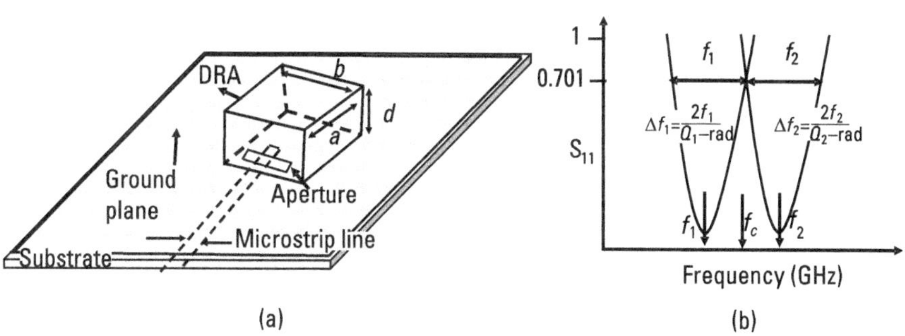

Figure 3.4 (a) Slot-coupled RDRA and (b) frequency response of the antenna [3, 4].

$$\frac{f_1}{Q_1}+\frac{f_2}{Q_2}=f_2-f_1 \tag{3.4}$$

Here f_1 and f_2 represent the resonant frequency and two resonating modes corresponding to Q_1 and Q_2, which are the unloaded radiation Q factors. The Q factor is inversely proportional to the impedance bandwidth of the antenna. Figure 3.4(b) depicts the frequency response of the antenna. A lower resonant frequency generates a –45° phase difference with respect to the center resonant frequency, and similarly a higher resonant frequency provides a –45° phase difference with respect to the center resonant frequency. Hence a total of 90° phase is generated corresponding to the resonating modes, which leads to circular polarization and thus improves the AR bandwidth.

In [5], researchers achieve circular polarization in a DRA through the microstrip feed technique. A CDRA is excited by an L-shaped microstrip line that is connected with a dual vertical microstrip feed line. This vertical feed arrangement excites $HE^x_{11\delta}$ and $HE^y_{11\delta}$ hybrid modes in the CDRA. Varying the height of the vertical microstrip feed line controls the quadrature (90°) phase shift of the exciting orthogonal modes. This antenna obtained a 30.37% (2.82–3.83-GHz) and 24.6% (2.75–3.52-GHz) impedance bandwidth and AR bandwidth, respectively. Throughout the operating band of the DRA, a 5.5-dBi average gain and a 96% radiation efficiency is achieved. This antenna prototype is fabricated using a dielectric material of permittivity 9.8. An unequal vertical feed line height creates a path difference between the vertical strip lines, which generates the phase difference in the resonating modes. The phase difference is calculated by

$$\textit{Phase difference} = {}^{2\pi}\!/_{\lambda} \times \textit{path difference} \tag{3.5}$$

Figure 3.5 shows the fabricated structure of the antenna and the E field distribution in the DRA corresponding to the vertical microstrip feed line.

In [6–9], circular polarization in various DRA geometries is achieved through a slot aperture–feeding technique, and the effects of the EM wave on polarization is detailed. In [6], an offset cross-slot-coupled CDRA is investigated for a wide bandwidth with circular polarization. The antenna obtained a 4.7% AR bandwidth for operation of the circular polarization. Two optimized unequal slots in each arm are incorporated in the ground plane, and the slot is situated at a 45° angle with respect to the microstrip line. The optimized first slot length and stub length at a corresponding frequency generates a –45° phase delay. Similarly the second slot length and stub length generates a +45° phase delay of E fields. Overall, the orthogonal fields' components are excited, which supports circular polarization radiation for wireless applications. The proposed

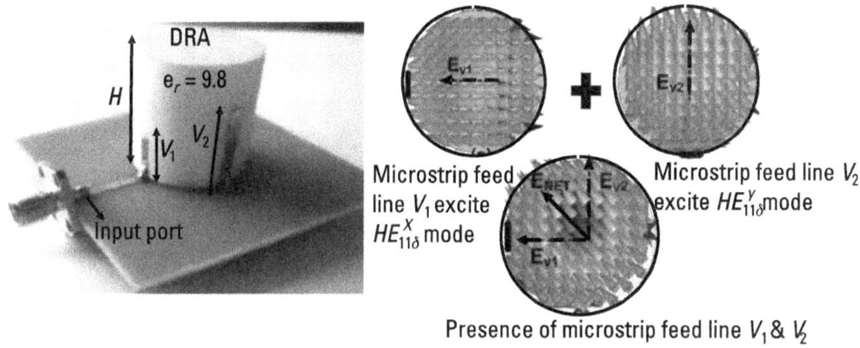

Figure 3.5 Fabricated structure of the antenna and E field distribution in the DRA corresponding to the vertical microstrip feed line [5].

antenna is constructed by Rogers TMM 10i laminate dielectric material whose dielectric constant is 9.8. Figure 3.6 shows a schematic diagram of the antenna.

A hemi-ellipsoidal DRA (HEDRA) excited by a cross-coupled aperture feeding is investigated for circular polarization in [7]. An impedance bandwidth ($S11 < -10$ dB) of 15.4%, and a 3-dB AR bandwidth of 3.9% are obtained. The geometry of the antenna and feeding methods results in degenerate orthogonal modes, which generates circular polarization in the DRA. In order to achieve circular polarization in [8], a partial ground plane with an F-shaped slot feed coupling is explored. Figure 3.7 illustrates the geometry of the proposed antenna. Circular polarization is obtained in the DRA by introducing an F-shaped slot in the partial ground plane to generate TE^x_{113} and TE^y_{113} orthogonal modes. These two orthogonal modes are responsible for enhancing the AR

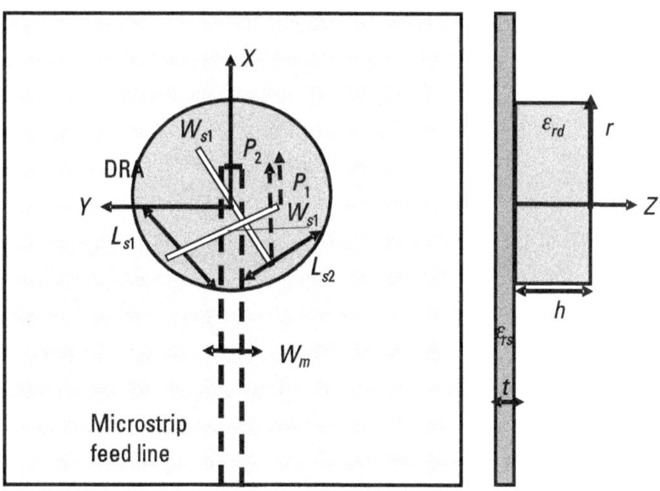

Figure 3.6 Schematic diagram of the antenna in [6].

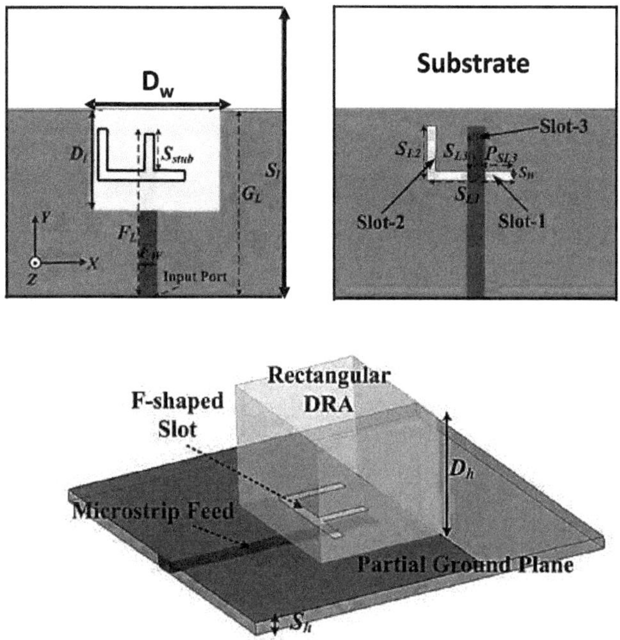

Figure 3.7 Geometry of the antenna proposed in [8]: (a) top view, (b) top view without DRA, and (c) 3-D structure.

bandwidth of the antenna. The F-slot, slot (S_{L1}) length, and width and length of the stub can be calculated as follows.

$$S_{L1} = \frac{0.4\lambda_0}{\sqrt{\varepsilon_{eff}}}, \varepsilon_{eff} = \frac{\varepsilon_{dr} + \varepsilon_{sub}}{2} \tag{3.6}$$

$$S_W = 0.2S_{L1} \text{ and } S_{stub} = \lambda_g/4 \tag{3.7}$$

In [9], a trapezoidal-shaped DRA is elaborated; the proposed antenna is excited by a a single 45° inclined slot fed by a single microstrip feed line and created on ground plane. A modified version (notch on top) of the trapezoid enhances the impedance bandwidth and AR bandwidth but does not improve the gain. A 21.5%, 3-dB AR bandwidth and a 34.6% (3.01–4.27-GHz) measured impedance bandwidth are obtained, respectively, with a measured gain from 5.29 to 8.39 dBi attained in the operating frequency range. For achieving the RHCP DRA, the DRA is rotated by (+45°) with respect to the rectangular slot. Similarly, a LHCP DRA can be obtained by making an angular displacement of (–45°). Figure 3.8 depicts a modified trapezoidal-shaped DRA with feeding techniques.

A cavity-backed circularly polarized DRA is excited by an asymmetrical U-slot in [10]. A source-independent resistive voltage source (RVS) is used to excite and generate the orthogonal field components. The finite-difference time domain (FDTD) method with a nonuniform orthogonal grid is used to analyze this new antenna configuration, with the feeding network modeled by an independent RVS source. In [11], a lumped resistively loaded monofilar spiral-slot is used to excite the rectangular DRA for wideband circular polarization. An 18.7% wide 3-dB AR bandwidth and a 53.5% impedance bandwidth are obtained for wireless application. Figure 3.9(a) presents the geometry of this DRA. This antenna design has two parts; first is the rectangular DR, and second is the lumped resistively loaded monofilar spiral slot that is used for excitation. It is constructed on the dielectric substrate and excited by a microstrip feed line. The antenna is fabricated with ceramic dielectric material ($\varepsilon_r = 12$). To minimize reflection at the end of spiral slot, a resistor is used with an impedance equal to the characteristic impedance of the slot line, and it is fixed on the slot arm. The magnetic current distribution corresponding to the slot [12] can be calculated by

$$\overrightarrow{M_s} = \overrightarrow{E_s} \times \hat{Z} \tag{3.8}$$

Here, $\overrightarrow{E_s}$ is the flow of the E field in the slot, and $\hat{Z}$ is the unit vector perpendicular to the ground plane. Figure. 3.9(b) depicts E field variation in different time steps. The resonant frequency corresponding TE^x_{111} and TE^y_{111} modes are excited, which enhances the AR bandwidth.

To excite the RDRA, an Archimedean spiral slot is used for achieving wideband circular polarization radiation in [13]. The Archimedean spiral slot structure seems like a broadband traveling-wave feeding structure. Two chip

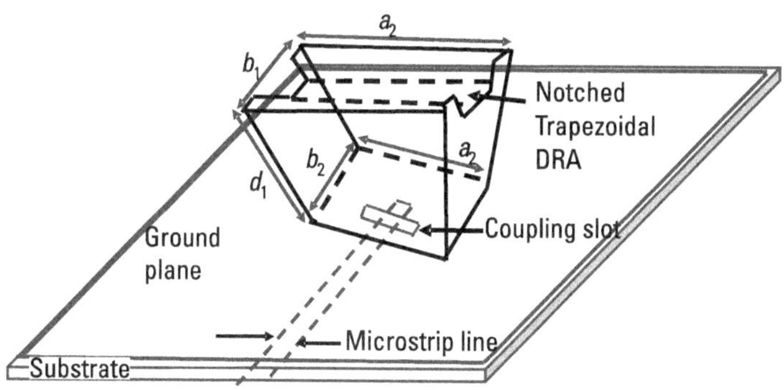

Figure 3.8 Modified trapezoidal-shaped DRA with feeding techniques [9].

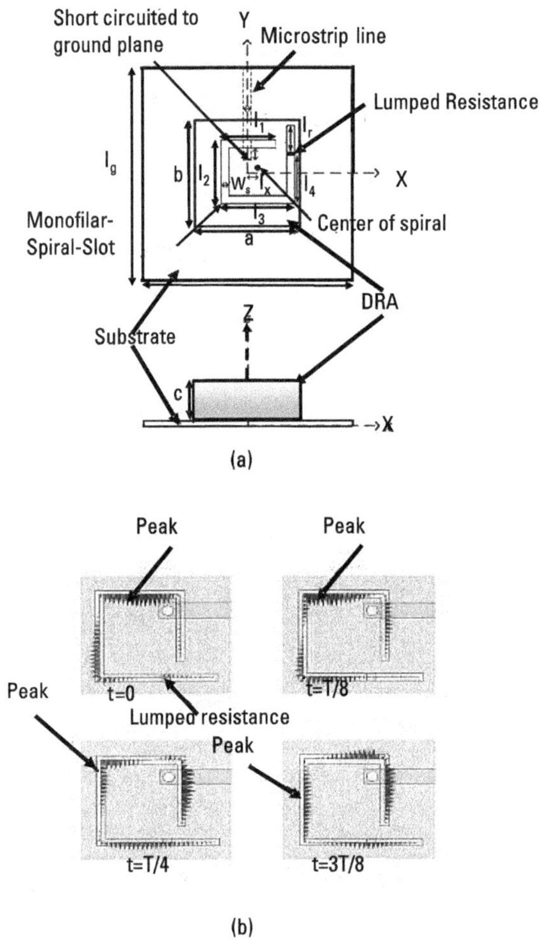

Figure 3.9 Circularly polarized configuration in a DRA: (a) geometry of the DRA and (b) electric field variation [11].

resistors are used at the end of the Archimedean spiral slot for matching of the impedance. The Archimedean spiral slot generates TE^x_{111} and TE^y_{111} orthogonal modes for circular polarization. Applying a parametric study on the spiral slots controls the polarization performance. Figure 3.10 shows the layout of the RDRA with the Archimedean spiral slot. The antenna is constructed from ceramic material with a dielectric permittivity $\varepsilon_r = 12$. 25.5% so that a wide 3-dB AR bandwidth is obtained.

Further techniques involving the introduction of parasitic elements in DRA to obtain circular polarization are discussed in [14–17]. In [14], a parasitic annular slot is used for achieving both linear polarization (LP) and CP in DRAs [14]. The HDRA is excited by an annular slot etched on the ground

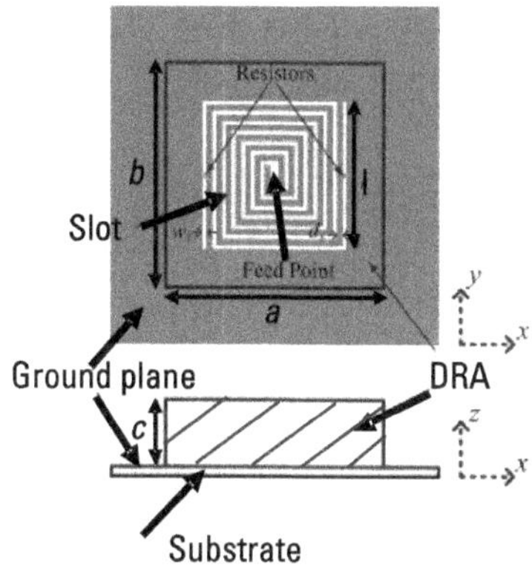

Figure 3.10 An RDRA with an Archimedean spiral slot excitation [13].

plane. By changing the radius of the parasitic slot, the antenna's resonant frequency can be controlled while maintaining good impedance bandwidth and a stable radiation pattern. Figure 3.11 shows the configuration of HDRA with excitation techniques. Moreover, in [15], to excite the CP in a DRA, rectangular parasitic metallic plates and pair of half rings DRs are incorporated as parasitic elements. A 52.2% (2.3–3.82-GHz) impedance bandwidth and a 46.8% (2.3–3.8-GHz) AR bandwidth are achieved. Similarly, in [16], a single parasitic patch is applied on an HDRA for CP. Conformal strip-fed techniques through HDRA are excited, and a fundamental TE_{111} mode is obtained.

In [17], an RDRA is excited by a single feeding method. A single feed is designed by an open half-loop antenna that consists of three metallic strips. The circularly polarized bandwidth of 7% is increased to 13% by further adding a concentric parasitic open half-loop in the structure, which is called the driven element. A 20% impedance-matching bandwidth is obtained through this antenna structure. The traveling current distribution achieves a 90° phase difference along the two half-loops for generating CP. Figure 3.12 presents an RDRA excited by a concentric open half-loops geometry configuration.

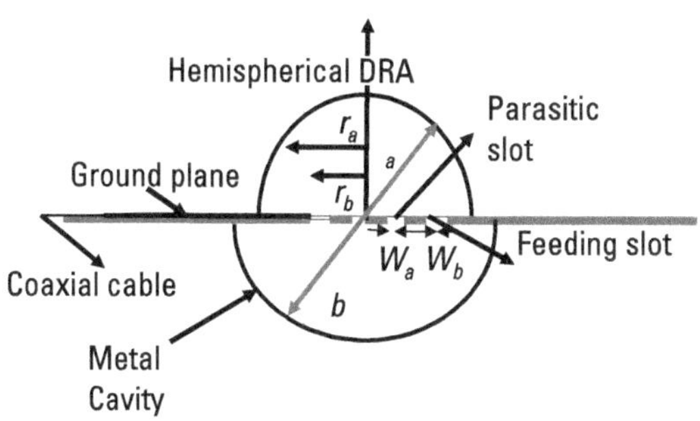

Figure 3.11 Configuration of an HDRA with excitation techniques [14].

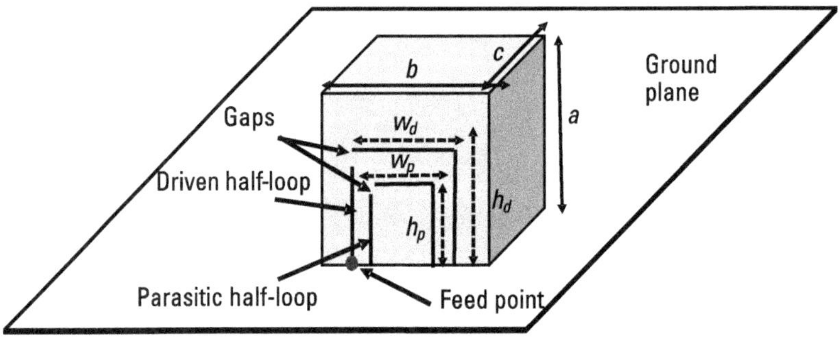

Figure 3.12 RDRA excited by a concentric open half-loop geometry [17].

Another approach to achieve CP in [18] is the conformal excitation technique in DRA. To obtain the desired AR bandwidth and impedance bandwidth, parameters of the spiral, like the spiral constant (*a*) and the winding constant, (ρ_0) are optimized. To construct the HDRA, $\varepsilon_r = 9.5$ dielectric material is selected with a radius of $r_1 = 1.25$ cm. The optimized parameters of the spiral arm obtained are $\rho_0 = 0.1415$ cm and $a = 0.0676$ cm, and the exciter wire radius is 0.25 mm for the best traveling current distribution in the DRA. A 12% impedance-matching bandwidth and 3.9% of a 3-dB AR bandwidth are obtained. Geometry of the DRA is excited by conformal wire as shown in Figure 3.13.

In [19], CP is achieved through an external helix exciter of the CDRA. A copper tape helix is connected by a coaxial line through a small section on the ground plane. This antenna feed network configuration offers a 3-dB AR bandwidth of 6.4 GHz. The helix feed network excites two orthogonal radiating modes in the phase quadrant, which contributes to generate a circularly polarized wave for communication. Figure 3.14 depicts the geometry of the proposed antenna with its feeding network.

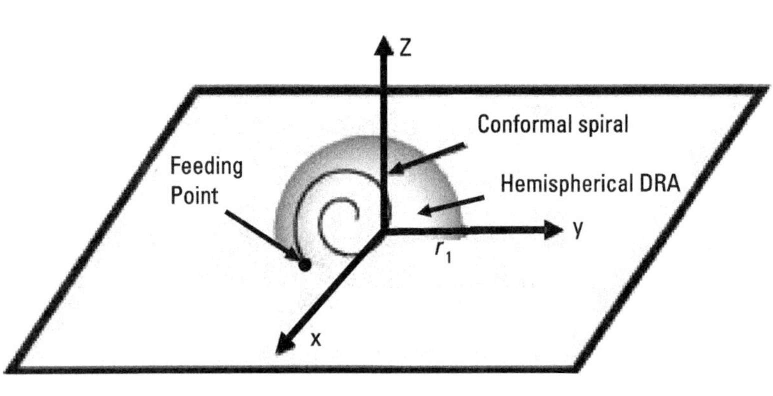

Figure 3.13 Geometry of a DRA excited by a spiral wire [18].

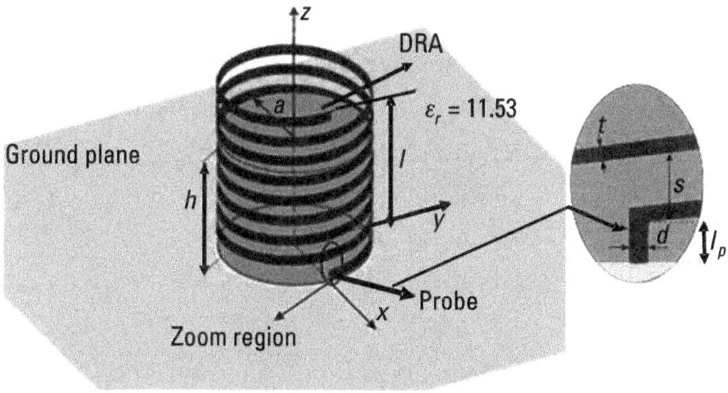

Figure 3.14 The geometry of the antenna with a feeding network proposed in [19].

In [20] and [21], respectively, the authors explore new techniques for CP, such as zonal-slot/DRA hybrid antenna without external ground plane and compact unilateral DRA. In [21], the author reports two sets of circularly polarized modes $TM_{10\delta}$ and $TE_{01\delta+1}$. This antenna is common in 2.4-GHz wireless local area network applications. In [22, 23], for CP, orthogonal modes are generated by feeding techniques and antenna performances enhanced by using a meta superstrate.

3.2 CP by Multiple-Feeding Techniques

In order to achieve CP, dual- or multiple-feeding techniques can also be used to excite the antenna. Two linearly polarized waves that are orthogonal and have equal amplitudes are capable of generating degenerate modes for CP. If DRA is excited with an identical feed point with an orthogonal feed position, CP can be generated. For exciting, two feeding signals (V_{x1} and V_{y1}) having equal amplitude with quadrature phase shift are required. Similar feeding techniques for obtaining the CP in DRA have involved a Wilkinson power divider, a hybrid coupler, a T junction, a conformal microstrip, a phase delay line, and printed microstrip [24]. Figure 3.15 shows various orthogonal feeding techniques.

Dual-point feeding techniques in DRA are used to excite E fields (E_{x1} and E_{y1}), which are dependent on excitation signals (V_{x1} and V_{y1}). To achieve good AR bandwidth, minimum mutual coupling between excitation ports is required. Dual-port feeding V_{x1} and V_{y1} input signals are applied with their receptive ports [25]. The AR can be calculated from (3.9) to (3.11). Let the excitation signal voltage be $V_{y1} = V_{x1}e^{j\Delta\Phi}$. The excited E field will be

$$E_{y1} = E_{x1}e^{-j3\Delta\Phi}\left(1 - \tilde{A}_0 e^{j\Delta\Phi} / 1 - \Gamma_0 e^{-j\Delta\Phi}\right) \tag{3.9}$$

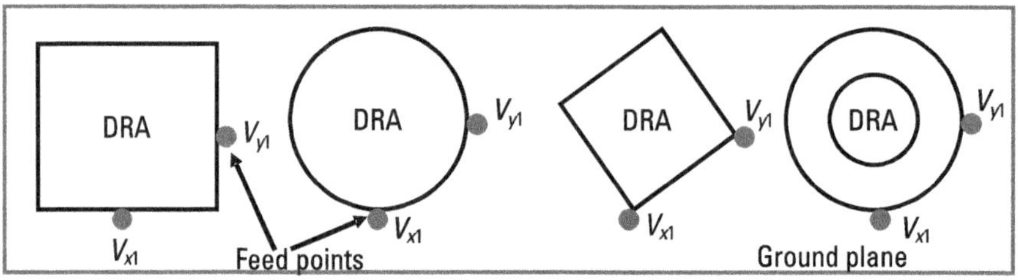

Figure 3.15 Various orthogonal feeding techniques.

where $\Delta\Phi$ is a phase difference between port-2 and port-1.

$$\Gamma_0 = \left(1 - S_{11}^2 + S_{12}^2\right) / 2S_{12}^2 \tag{3.10}$$

In (3.10), Γ_0 is the reflection coefficient, S_{11} is the reflection coefficient of port-1, and S_{12} is the transmission coefficient between port-2 and port-1. E_{x1} and E_{y1} are the field components that are generated by port-1 and port-2, respectively. The AR is calculated by [1]

$$\textit{Axial Ratio} = 10\log\left[{}^{1+|\Gamma_0|}\Big/_{1-|\Gamma_0|}\right] \tag{3.11}$$

It is observed that if coupling between two feeding ports improves, the AR bandwidth performance degrades.

Dual- or multiple-feeding techniques are inspected for achieving CP in [26–40]. Several dual coaxial probe-feeding techniques are described in [26–28] to generate CP. Circular polarized CDRA [26] is constructed with a high dielectric material ($\varepsilon_r = 36.2$). The antenna is efficiently excited by the dual coaxial probes, which are located at the center and outer edge of the cylinder; 3-dB couplers are used to give the excitation of the coaxial probe, which generates a phase difference in field components. One input port acts as a feeding port, and another port is isolated. This antenna structure is placed on a copper ground plane. The hybrid $HE_{11\delta}$ mode is easily excited in a DRA when the feed is given through the vertical coaxial probe conductor. Due to the antenna excitation configuration, two mutually orthogonal and degenerate $HE_{11\delta}$ modes are generated of equal amplitude and with a quadrature phase difference. A 3-dB quadrature coupler is used to excite the antenna, generating orthogonal field components as shown in Figure 3.16.

Similarly, in [27], two orthogonal $HE_{11\delta}$ modes are generated by two coaxial probe feeds and soldered with a microstrip feed line. This microstrip feed line is printed on an RT Duroid 5880 substrate with a dielectric constant $\varepsilon_r = 2.2$. A quadrature (90°) phase shift between the two excitation signals

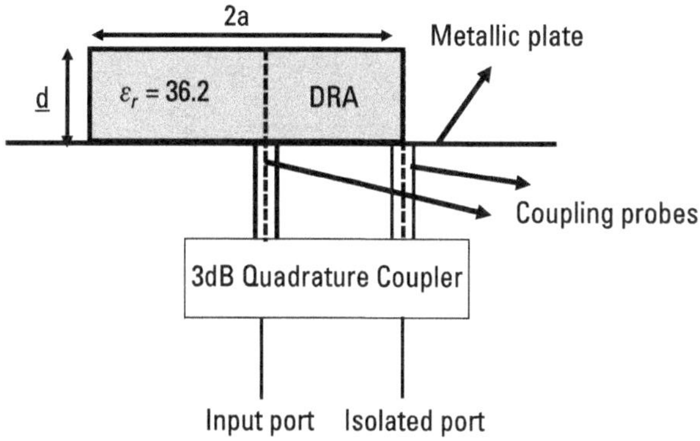

Figure 3.16 Configuration of a circular polarized DRA with feeding techniques [26].

is achieved through a semicircle-shaped $\lambda/4$ microstrip transmission line. The antenna offers a 3.9% impedance bandwidth with a 0.45-dB AR in boresight. Figure 3.17 depicts the antenna geometry with microstrip feeding techniques.

CP is achieved in a magneto DRA through a dual-probe feeding network in [28]. A 90° hybrid coupler is used to generate orthogonal field components with a quadrature phase difference. This coupler divides the input power equally. An operating bandwidth of more than 50% is obtained for a 240–420-MHz frequency range application. Further, dual orthogonal linear polarization is achieved with the conformal-dual DRAs [29]. Good CP excitation is obtained by the two conformal microstrip lines. Figure 3.18 demonstrates the geometry of the proposed antenna and its feeding technique. The rectangular-shaped DRA is designed with ceramic material with a dielectric permittivity of $\varepsilon_r = 79$. A microstrip feed line is properly selected to achieve good impedance matching. Two conformal microstrip lines are placed on the surface of the DRA in the orthogonal direction, and the length of the conformal strip line is optimized for obtaining the AR. The antenna offers a 58-MHz frequency band for various wireless applications. If the distance of the two feeding ports increases then port isolation is improved.

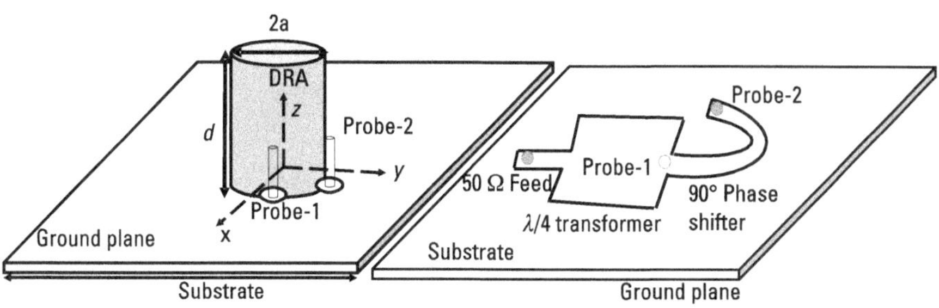

Figure 3.17 Antenna geometry with microstrip feeding techniques [27].

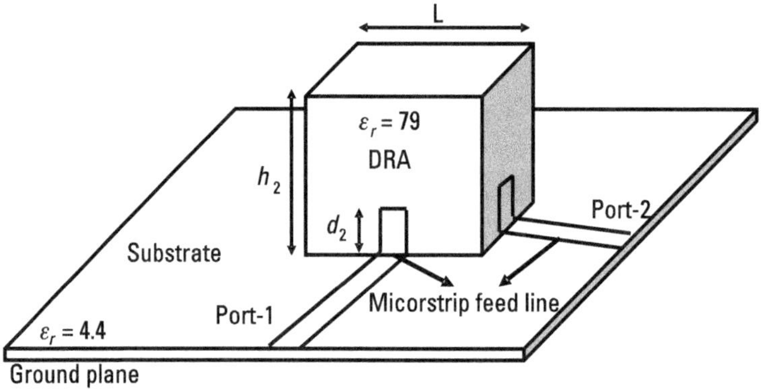

Figure 3.18 Geometry of antenna and feeding technique proposed in [29].

A dual-feed $HE_{11\delta}$ mode DRA-loaded monopole antenna is examined in [30] for mobile communication terminals. The orthogonal feed position of the DRA and the integrated monopole of the DRA are capable of generating orthogonal field components. Similarly in [31], a dual-mode quadrature-fed wideband circularly polarized DRA is investigated for achieving CP. The quadrature-feeding technique is one of the convenient techniques for obtaining CP in a DRA. An RDRA structure is fed by two vertical conductor strips. TE_{111} and TE_{113} modes contribute to generate a wideband CP. A vertical feed strip (V_1) excites the TE^y_{111} and TE^y_{113} modes; similarly a vertical strip (V_2) excites the TE^x_{111} and TE^x_{113} higher-order mode. A combination of these resonant modes achieves a wide CP bandwidth. The antenna attains a 32.8% (2.7–3.76-GHz) impedance bandwidth and a 2.6–3.78-GHz AR bandwidth for various wireless applications. Figure 3.19 illustrates the antenna geometry of the quadrature-fed DRA.

A wideband, dual-feed, dual-sense, circularly polarized DRA is investigated in [32]. The antenna is excited by an L-shaped slot aperture with a microstrip

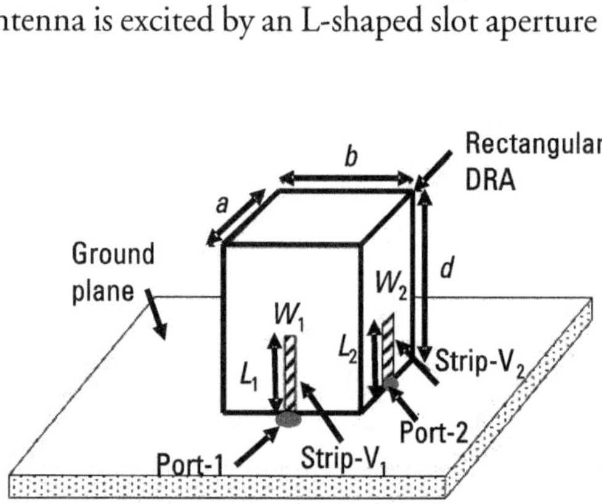

Figure 3.19 Antenna geometry of the quadrature-fed DRA detailed in [31].

feed line and a vertical metal strip. Using this feeding technique, a dual-band dual-sense CP triangular DRA is achieved. A proposed bow tie-shaped DRA consists of two triangular-shaped dielectric slabs. The antenna is fabricated and experimentally verified. A 63.7% wide 10-dB return loss bandwidth and a 27.1% left-hand and 12.8% right-hand broad 3-dB ARBW are obtained by the antenna. TE^x_{111} and TE^y_{111} modes are excited in the lower sideband, and TE^x_{151} and TE^y_{151} higher-order modes are excited in the upper sideband. Figure 3.20 shows this DRA configuration and its feeding methods.

Researchers have explored some new feeding techniques [33–37] for achieving CP in DRA. CP is achieved by dual conformal strips in a DRA [33]. This antenna structure offers a more than 20% impedance bandwidth. For the excitation of input signals in the DRA, a hybrid coupler is used. The hybrid coupler generates orthogonal signals with a quadrature phase shift in two input signals. Similarly in [34], a quadruple strip-feed CDRA is examined for CP. The antenna offers a 34.5% (1.75–2.48-GHz) impedance bandwidth (S_{11} < 10 dB) and a 25.9% (1.65–2.14-GHz) AR bandwidth (AR 3-dB) for wireless applications. In this case, a pair of 90° hybrid couplers is used. To improve the energy coupling between the microstrip feed line to the DRA, quadruple vertical conformal strips are connected around the circumference of the DRA. A pair of 90° hybrid couplers provides four vertical strips in equal amount of power and excitation phases of 0°, 90°, 180°, and 270°, respectively. Figure 3.21 depicts the configuration of the circularly polarized quadruple strip-fed CDRA with the 90° hybrid coupler pair.

In [35], dual underlaid hybrid couplers are used for exciting the DRA. The proposed antenna offers a 27.7% 3-dB AR bandwidth for wideband applications. In [36], another wideband circularly polarized DRA is investigated with an underlaid quadrature coupler. The underlaid coupler is situated inside the hollow region of the DRA. The DRA's loaded vertical strip offers the matched load for the coupler. In this technique, a 10.54% wide AR bandwidth is obtained. A quadrature coupler is printed on a substrate with a dielectric constant of ε_r = 6.15. A pair of conducting vertical strips is fixed on two adjacent side

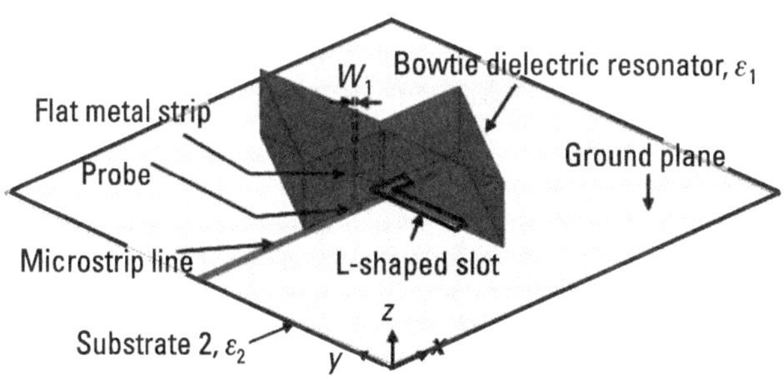

Figure 3.20 Configuration of the DRA and feeding methods in [32].

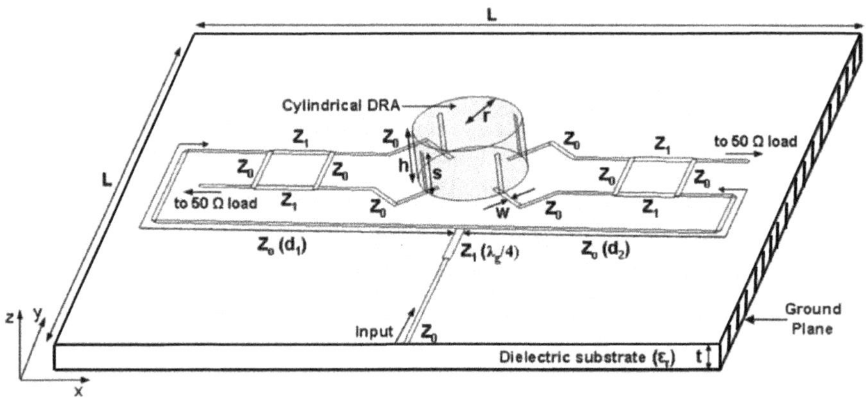

Figure 3.21 Configuration of the circularly polarized quadruple strip-fed CDRA with the 90° hybrid coupler pair.

walls of the DRA to generate the degenerate TE^{x}_{111} and TE^{y}_{111} orthogonal modes. Two vertical conducting strips are integrated to the 0° and 90° output ports of the coupler for achieving quadrature inputs. Figure 3.22 shows this proposed DRA with its underlaid quadrature coupler and input electric field excitation.

In [38], in addition to other feeding techniques, a new hybrid DRA is investigated for wideband CP, using a switch line coupler for excitation of the DRA. A switched line coupler offers good impedance matching, equal power distribution, and a 90° phase shift of the output ports of line coupler. The proposed antenna obtained a 47.69%, 3-dB AR bandwidth and 48.46% impedance-matching bandwidth for wideband applications. Next, [40] considers a low-profile DRA for circular polarization. The proposed antenna is constructed out of a T-shaped dielectric resonator and a pair of microstrip-fed stubs printed on a substrate with a truncated ground plane for providing the suitable feed excitation. Figure 3.23 illustrates this fabricated antenna prototype. For achieving CP, multiple orthogonal modes are generated. The antenna offers a 17.2% (4.98–5.92-GHz) bandwidth and a 17.9% (5.02–6.03-GHz) AR bandwidth. Peak gains of 3.2 and 3.3 dB are attained at two feed ports, respectively.

3.3 CP by Geometry Modification Techniques

This section discusses various modified geometries such as modified shapes on DRA geometry [41, 42], arrangements of stacked dielectric material [44–59], a modified ground plane with a reformed antenna geometry [60–64], and metallic patch loading [65–69] in DRA. Figure 3.24 illustrates several modified DRA geometries for obtaining CP.

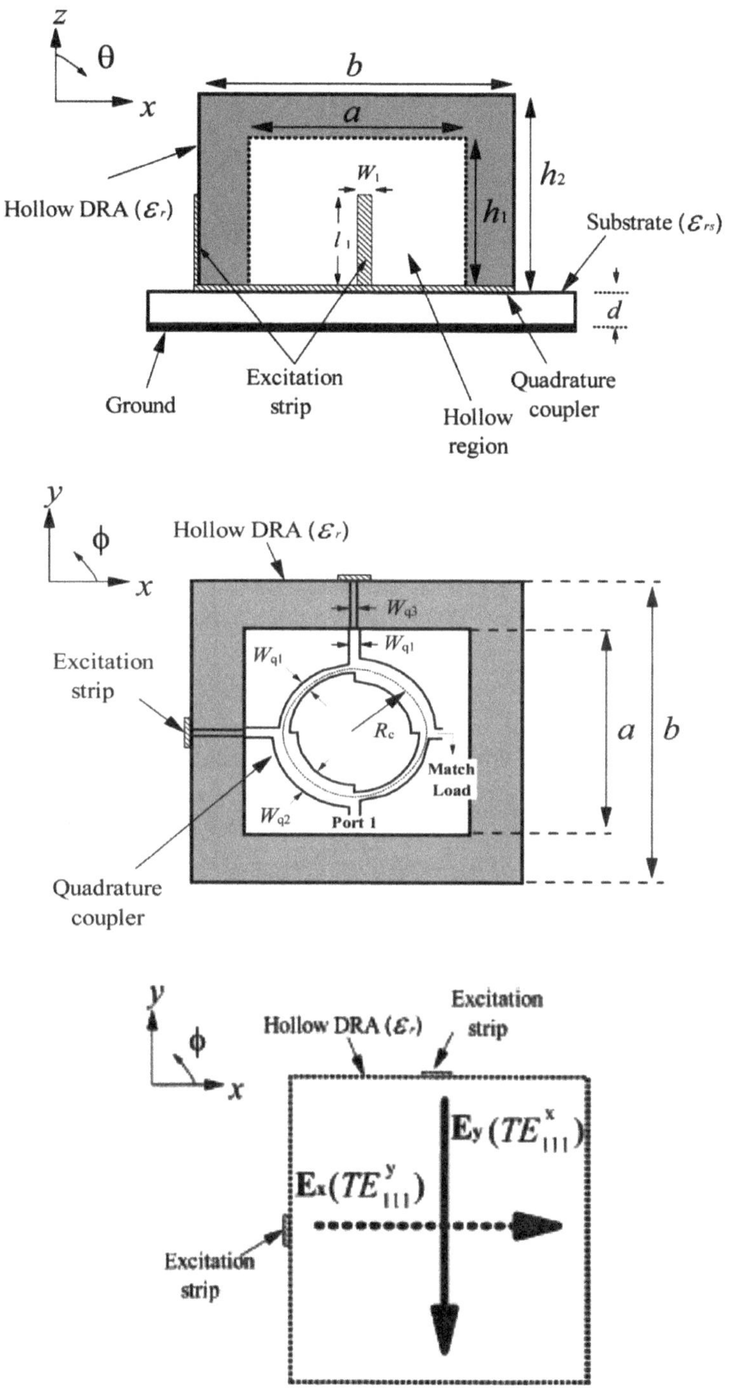

Figure 3.22 The DRA with an underlaid quadrature coupler and input electric field excitation proposed in [36].

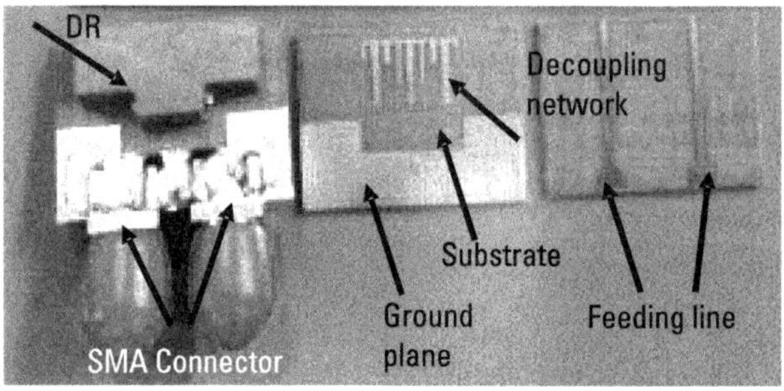

Figure 3.23 Fabricated antenna prototype [40].

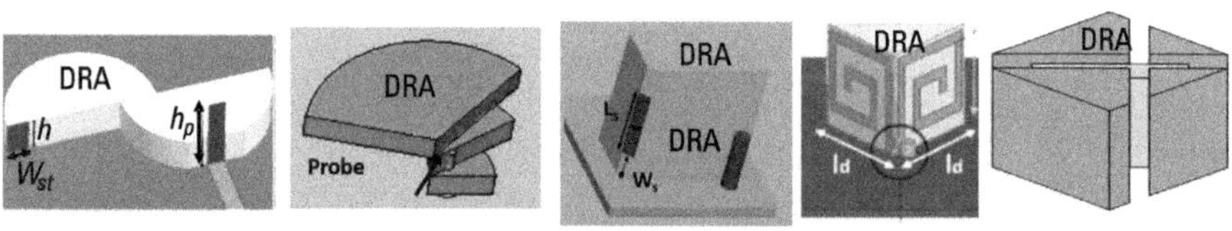

Figure 3.24 Several modified DRA geometries.

In [41, 42], a circular sector–shaped CDRA [41] and a semi-eccentric annular-shaped DRA [42] are reported for obtaining CP. Optimum antenna design parameters like the radius-to-height ratio and feed position of the circular sector of the DRA excite two resonant modes, which are orthogonal and in phase quadrature for CP. This antenna is excited through a single-probe coaxial feeding network, and exciting modes behavior is analyzed with two different feeding positions, at the right corner and center of the DRA. If the feeding network is located in the right corner position, the $TM_{11\delta}$ mode is excited. When the feeding network is at the center position, the $TM_{21\delta}$ mode is excited. At the resonant frequency, a lower resonant mode is driven at –45°, and the upper resonant mode is driven at +45°; a total of 90° in phase difference appears between the two exciting modes. Both f_1 and f_2 are resonant frequencies of orthogonal modes, and a quadrature phase difference appears in the two modes when the following condition is fulfilled by the resonant frequency:

$$f_1 + {\Delta f_1}/{2} = f_2 - {\Delta f_2}/{2} \tag{3.12}$$

where Δf is the 3-dB bandwidth of $|S_{11}|$. Thus, these radiating modes contribute to achieve CP. Figure 3.25 shows the geometry of the proposed antenna. The antenna is constructed by machining an Eccostock HiK dielectric rod with a dielectric constant of ε_r=12. The sector CDRA is placed on the copper ground plane, and a 10% operating bandwidth is obtained for wireless communication.

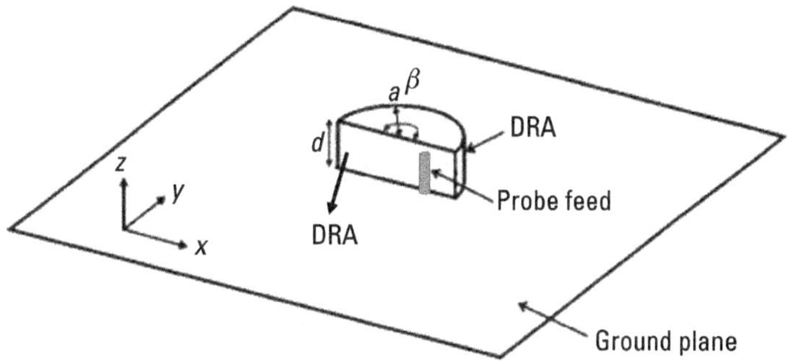

Figure 3.25 The geometry of the proposed antenna in [42].

In [43], different forms of circular-sectored DRAs (CS-DRAs) formed with different sector angles (β) are analyzed for improving the circularly polarized performance. The cylindrical DRA aspect ratio = $2H/r$, sector angle (β), and excitation feed position are optimized to obtain a wide AR bandwidth. Two fundamental mode $TM_{\upsilon,1,\delta}$ and $TM_{2\upsilon,1,\delta}$ resonating modes are excited to achieve orthogonal polarizations. The antenna is made with alumina dielectric material ($\varepsilon_r = 9.8$) and placed over the 50×50 mm^2 FR-4 epoxy substrate; the copper layer of the ground plane is below the substrate. This antenna configuration offers 15.2% (5.47–6.37-GHz) and 15.46% (5.43–6.34-GHz) simulated and measured AR bandwidths, and 58.5% (4.24–7.75 GHz) and 48.3% (4.65–7.6-GHz) simulated and measured reflection coefficient bandwidths. Figure 3.26 shows the sectored DRA with a variety of sector angles (β), aspect ratios $\alpha = 2H/r$, and feed positions.

Next, we investigate antennas to achieve wideband CP: RDRAs using square-shaped slots [44], spidron fractal DRAs [45], and inverted-sigmoid-shaped

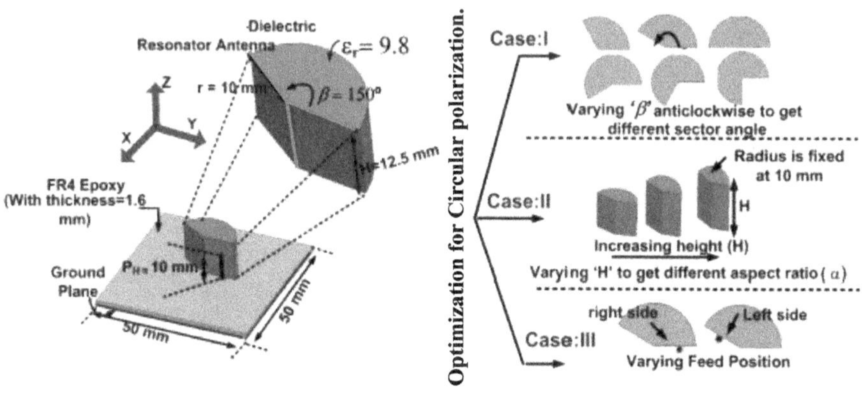

Figure 3.26 CDRA for CP: (a) a sectored DRA and (b) varying the sector angle (β), the aspect ratio α, and the feed position.

multiband DRAs [46]. In [47], a Fibonacci series approach-based CDRA is examined. For achieving a 3-dB AR bandwidth, the generation of two orthogonal degenerate modes is dependent on the proper selection of the cylinder aspect ratio (*a*/*h*), dielectric material (ε_r), and feed location in the antenna. Orthogonal degenerate hybrid modes $HE_{11\delta}$ and $HE_{12\delta}$ are excited by the proposed antenna. A Fibonacci CDRA is designed through Fibonacci numbers form of the radii of a quarter segment of a CDRA, which is then joined to form the proposed DRA structure. This antenna configuration is fabricated using FR-4 dielectric material (ε_r= 4.3). Figure 3.27 depicts this antenna's design configuration and a fabricated prototype of the antenna. The proposed antenna covers a 106.5% impedance bandwidth and a 18.9%, 3-dB axial bandwidth for satellite communication, navigation, and other modern wireless applications system.

In [48], to obtain CP, four rectangular dielectric layers are rotated at an ($\beta = 30°$) angle relative to each other and are stacked over each other. This arrangement is excited through aperture coupling. The generated fields in the inclined dielectric layers have both components of $\widehat{a_x}$ and $\widehat{a_y}$, and the rotation distances between the dielectric layers lead to a phase difference between the adjacent layers. CP is achieved by the proper selection of the dielectric layer rotation angle (β) and the layer spacing. CP is obtained due to the excitation of two orthogonal modes, TE^x_{111} and TE^y_{111}, in the antenna. This antenna geometry provides 6% from a 9.55–10.15-GHz AR bandwidth. Figure 3.28 shows the fabricated rotated four rectangular dielectric-layer antenna structure.

In [50], a stair-shaped DRA [50] is discussed for WLAN applications. For achieving CP, two rectangular dielectric layers are joined together to form a stair-shaped DRA. The excitation of multiple orthogonal modes in the DRA leads to an enhanced CP bandwidth. Placing the two dielectric layers side by side to form a stair-shaped DRA encourages the generation of two TE^x_{121} and TE^y_{211} modes. Figure 3.29 shows this antenna's geometry configuration. A 22% 3-dB AR bandwidth and a 37% impedance bandwidth is obtained by the antenna. In addition, it provides a stable broadside radiation with a gain range of 4.5–5.7 dB in the bandwidth.

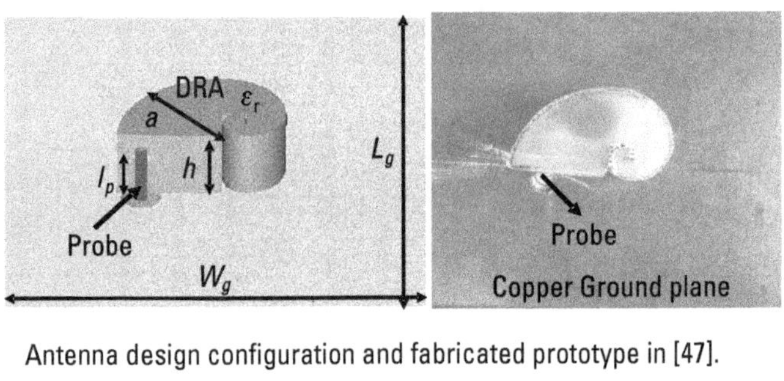

Figure 3.27 Antenna design configuration and fabricated prototype in [47].

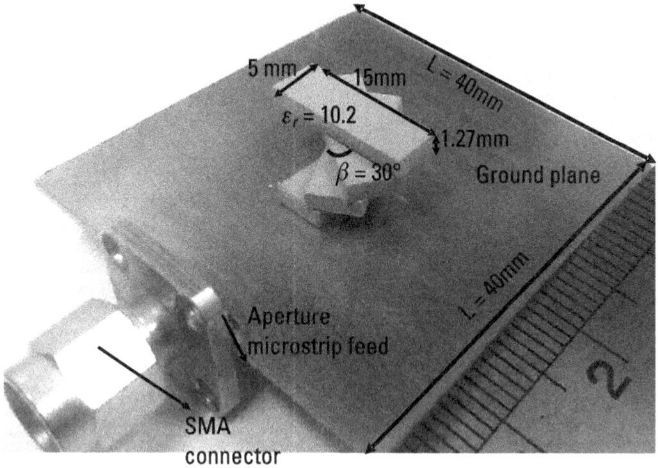

Figure 3.28 Fabricated rotated four rectangular dielectric-layer antenna structure [48].

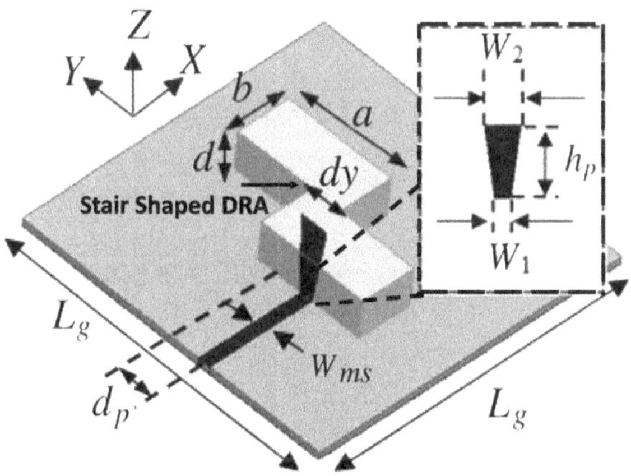

Figure 3.29 Antenna geometry configuration [50].

In [52], a segmented DRA is designed for an ISM band application. To construct the proposed antenna, three different (120°, 60°, and 30°) segments of different cylindrical radii (r_1, r_2, and r_3) with stacked angular displacement are used for CP. Proper segmentation of cylinder stacked with specific angular displacement offers a quadrature phase shift of the orthogonal field components. The simulated and measured impedance bandwidths of the proposed antenna obtained are 90% (3.3–8.7 GHz) and 83.4% (3.5–8.5 GHz). A similar measured AR bandwidth is 53.8% (3.8–6.6 GHz) and 58.5% (3.5–6.4 GHz). The E fields of the middle layer generate a 45° phase deviation with respect to the lower layer, and the E fields of the upper layer generate a 45° phase deviation with respect to the middle layer. This creates a total of 90° of phase deviation for the generation of orthogonal E field components. Figure 3.30 depicts the electric field variations at 6-GHz frequency.

Furthermore, in [54], a modified geometry with a parasitic strip is used to obtain CP. The parasitic strips excite a CP mode, and this mode combines with the DRA mode to achieve a more than 25% wideband AR bandwidth. Figure 3.31 shows the configuration of this proposed antenna.

Several researchers propose DRAs with modified ground planes and various feeding techniques to explore the CP property of antennas in [59–63]. An elliptical dielectric ring resonator with an aperture-coupled microstrip patch reversed T-shaped slot [59], a rectangular ring dielectric resonator and parasitic printed loops [60], a sectored conical dielectric resonator antenna excited by an S-shaped slot [62], and a cubic-shaped DRA [63] are investigated for wideband CP. In [63], a new question-mark-shaped microstrip feeding network technique is used to excite the antenna and two orthogonal modes $TE_{11\delta}$ with a quadrature phase difference. Figure 3.32 presents the geometry of the proposed antenna. The dielectric resonator is designed using Al2O3 dielectric material with dielectric permittivity of $\varepsilon_r = 9.8$ and a loss tangent of $\delta = 0.002$. To achieve the best antenna performance, the effect of the ground plane and various dielectric materials are studied on the bandwidth and AR performance. If the dimension of the ground plane is increased, the input impedance bandwidth decreases because of increased coupling between the ground plane and the question-mark-shaped feeding network. This antenna geometry configuration offers a 10-dB impedance bandwidth of 35.35% and a 3-dB AR bandwidth of 20.62% in the broadside direction. This antenna is suitable for WiMAX (3.3–3.7-GHz) applications.

Patch-loading techniques are used in DRAs to attain CP in [64–68]. In [64], a top-loaded modified Alford loop is used in a DRA for achieving CP. Figure 3.33 illustrates the configuration of the proposed antenna geometry, which is fed by a coaxial probe feed technique. Two orthogonally polarized field components are equal in amplit ude but different in phase by 90°; two degenerate

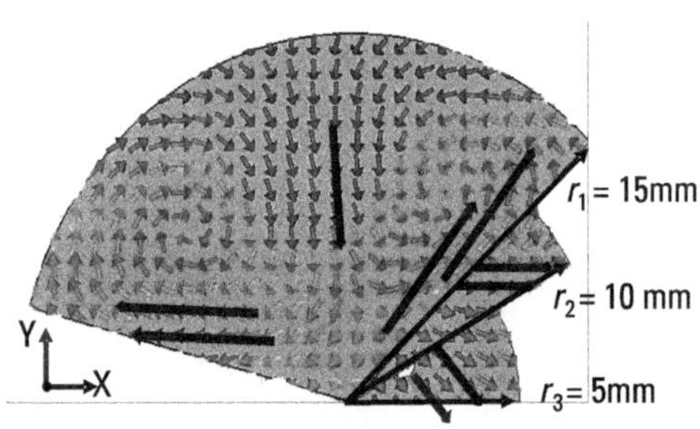

Figure 3.30 E field variations at a 6-GHz frequency of segmented stacked DRA [52].

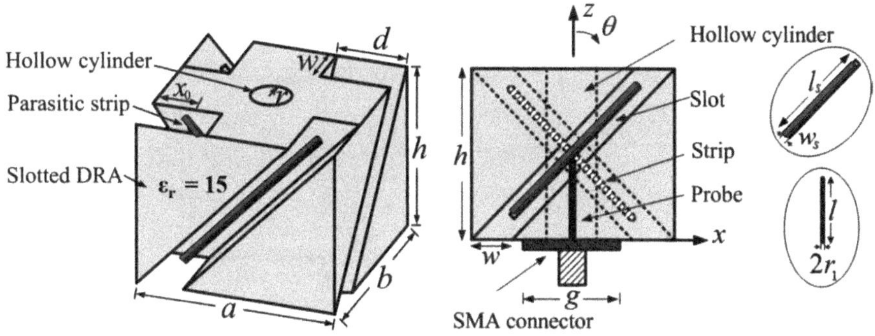

Figure 3.31 Configuration of the antenna proposed in [54].

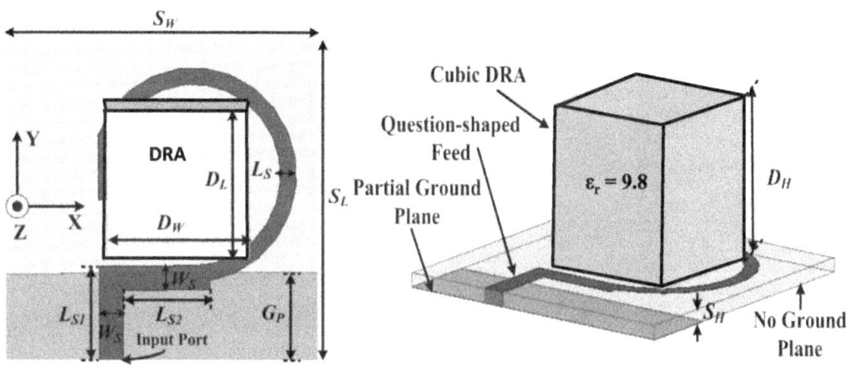

Figure 3.32 Geometry of the antenna proposed in [63].

$HE^x_{12\delta+1}$ and $HE^y_{12\delta+1}$ modes of the DRA are excited in a broadside CP radiation direction by a modified Alford patch loop.

Similarly, in [66], a compact equilateral triangular DRA (TDRA) is designed for a CP application. Coupling based on a dual-facet spiral loop is incorporated on two sides of the surface of an equilateral triangular dielectric to improve the AR performance. Hence it creates phase excitation on two side walls of the equilateral TDRA to conform the CP at 7.5- and 8.7-GHz frequency. Figure 3.34 shows a schematic diagram of the proposed antenna, which consists of an equilateral triangular dielectric resonator of edge (l_d) = 12 mm, height (h_d) = 10 mm, and $\varepsilon_r = 10$ low-loss dielectric material. It is placed on a copper ground plane with dimensions of $l_g = w_g = 50$ mm, and the coaxial probe is coupled with an SMA connector in the ground plane. Two rectangular spiral loops are designed from seven individually copper cut strips. For fabrication, an adhesive copper tape is used to construct the spiral loops.

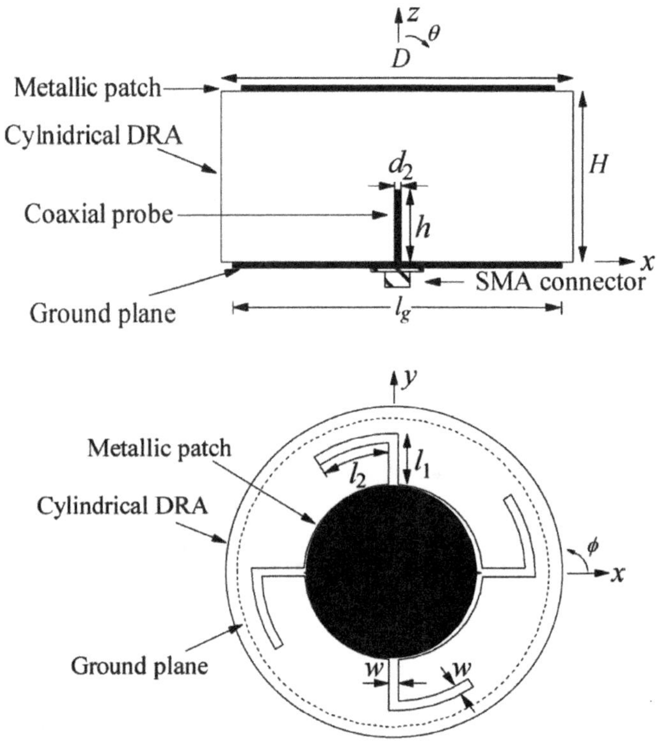

Figure 3.33 Configuration of the proposed antenna geometry in [64].

3.4 CP by the Array Configuration Technique

An antenna array consists of multiple similar types of antenna elements, which are arranged in a special way to improve the performance of the antenna parameters. In an antenna array, the spacing between the elements and feeding methods plays an important role in enhancing the directivity, radiation pattern, bandwidth, CP, gain, and signal-to-noise ratio. Usually a single element of an antenna offers a low gain and a wide beamwidth radiation pattern with a narrow CP band for long-distance communication and space communication. Enlarging or increasing the number of antenna elements leads to improved radiation characteristics. Figure 3.35 depicts various antenna array configurations.

A variety of feeding network arrangements is applied in the DRA array for generating phase delay or quadrature phase and orthogonal modes for CP. For array application, feeding phase excitation for an antenna can be achieved through passive phase delay (microstrip-line path length). If a two-antenna array system is excited by a series or parallel microstrip feed line, the spacing between antenna elements d_s and (Φ_{s1}, Φ_{s2}) is exciting signal phases of antenna

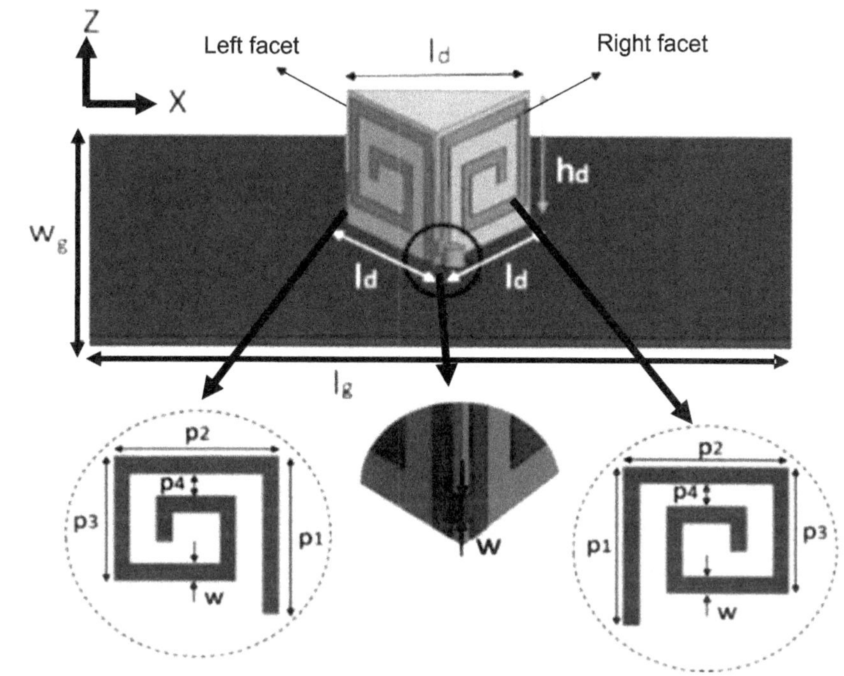

Figure 3.34 Schematic diagram of the proposed TDRA antenna [66].

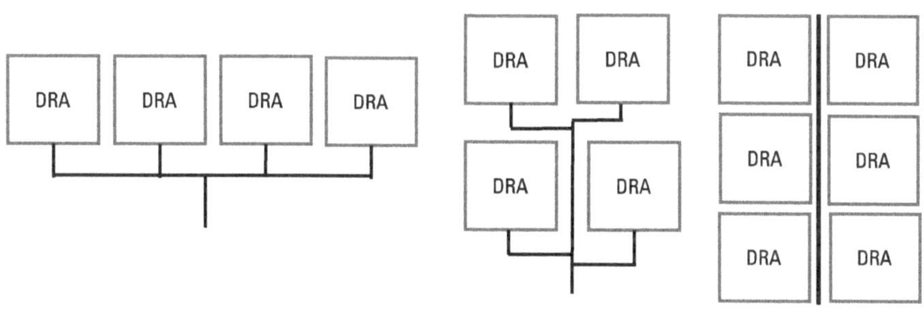

Figure 3.35 Geometry of array antenna arrangements.

elements. In antenna exciting feeding the signal phase delay depends on the path length of the feeding network. For a microstrip feeding network, the desired excitation phase in the antenna is achieved by adjusting the distance (d_s) between the antenna elements [69]. For series and parallel microstrip feed networks, the phase is generated as given in (3.13) and (3.14).

For a series feeding network,

$$\frac{d_s + 2l}{\lambda_g} = \frac{\varphi_{s2} - \varphi_{s1}}{2\pi} \tag{3.13}$$

For a parallel feeding network,

$$\frac{d_s - 2l}{\lambda_g} = \frac{\varphi_{s2} - \varphi_{s1}}{2\pi} \tag{3.14}$$

In (3.13) and (3.14), λ_g is the guided wavelength of the microstrip line, (Φ_{s1}, Φ_{s2}) is the exciting phase of the antenna elements, and d_s is the spacing between the antenna elements. Figure 3.36 shows a series and parallel microstrip feeding network.

In antennas, to generate CP, the exciting orthogonal modes and quadrature phase are dependent on the feeding network and the antenna element arrangements. This section covers various array configurations of DRAs for achieving CP [70–80]. A pair of dielectric resonator antennas is designed for broadband CP in [70]. A new 4 × 4-element DRA planar array having uniform aperture distribution is constructed with four DRA elements of subarrays. A 15% AR bandwidth below 2 dB was obtained from this new type of DRA array configuration. Figure 3.37 illustrates the DRA pair unit and subarray antenna configuration and the feed network. Due to the microstrip feed configuration, a pair of unit DRAs generates a $TM_{2v,1,\delta}$ phase delay in exciting the field components that contributed to the CP. Each element of a DRA pair-unit is excited uniformly by power distribution and differential phase ($F_1 = F_1 TM_{2v,1,\delta}$), which is controlled by their orthogonal feed points F_1 and F_2.

Next, a microstrip with aperture coupling is used for generating quadrature phase in DRA elements in [71–74]. A 2 × 2 cross-shaped DRA [71] is investigated for circular polarization. The DRA array is excited through the sequential rotation of the aperture microstrip feed, and the CP bandwidth is improved from 5% to 16% as compared to that of a single-element DRA. The DRA array offers a 25% wide impedance bandwidth for wireless applications. However, the antenna consists of low-loss dielectric material (Rogers 6010) having a dielectric permittivity of $\varepsilon_r = 10.8$. The arms of the DRA are rotated by 45° to the length of the aperture slot. Hence each arm is excited by equal

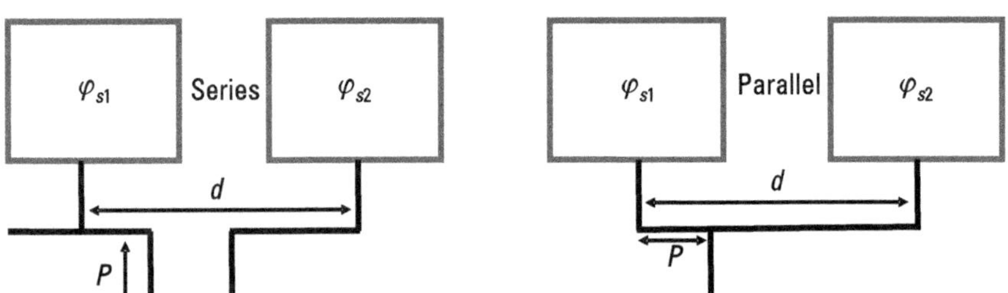

Figure 3.36 Series and parallel microstrip feeding network.

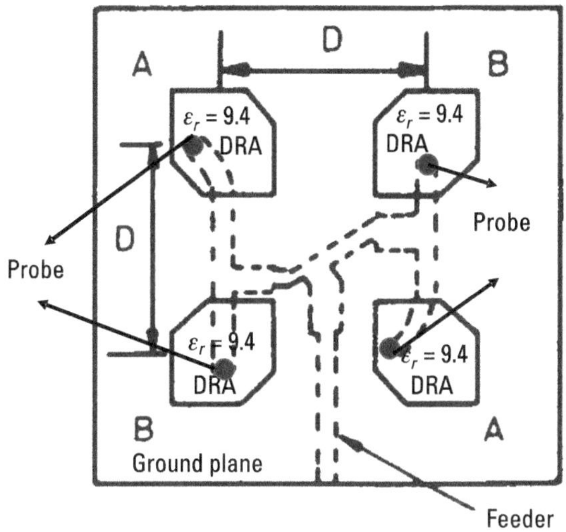

Figure 3.37 DRA pair unit and subarray antenna configuration and feeding network [70].

magnitude. Also, to obtain CP, lengths of the cross DRA were optimized to resonate in phase quadrature at resonant frequency.

Figure 3.38 shows an array of cross DRA excited by a sequential feeding network. The slot feed points are on the circumference of a circle on the substrate. Their slot lengths are oriented tangential to the circumference of circle. In order to minimize the mutual coupling effect in the feeding point, a microstrip feed line is located below the substrate, and it has less effect on the radiation pattern of the array. Sequentially rotated feeding aperture lines generate 0°, 90°, 180°, and 270° quadrature phase for achieving the AR.

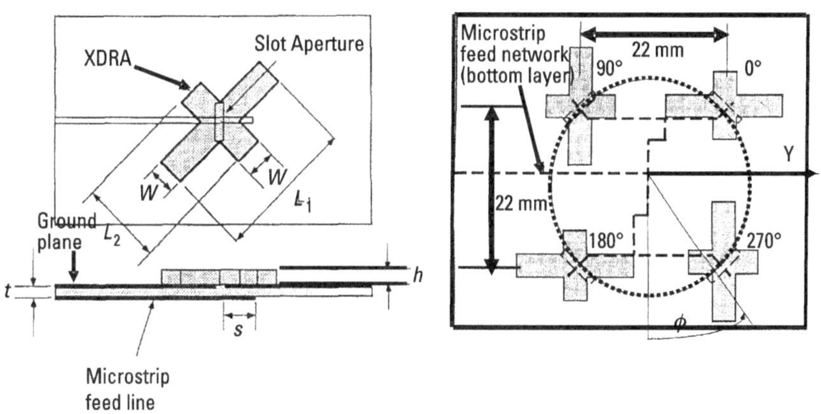

Figure 3.38 An array of cross DRA excited by a sequential feeding network [71].

Similarly, a sequential rotation technique is used for the design of circularly polarized DRA subarrays in [72]. The maximum gain of the CP-element subarray obtained was 12 dBi with a 19% impedance bandwidth. In [73], a 2 × 2 rectangular DRA array with metallic printed loading was investigated for achieving wideband CP. Figure 3.39 shows the proposed antenna structure [73]. The antenna array consists of alumina ($\varepsilon_r = 9.9$) low-loss dielectric material. The metallic strip is diagonally printed on the surface of the square-base dielectric resonator. For CP, the proposed configuration of the DRA is therefore the excitation of two orthogonal modes such as TE^x_{111} and TE^y_{111} with the same amplitude and phase quadrature. Moreover, the impedance-matching bandwidth and AR bandwidth can be controlled by adjusting the slot length (l_s), width (w_s), and stub length (s).

A four-element perturbed rectangular dielectric resonator excited by a coaxial probe is presented in [74]. Feeding arrangements and excitations for DRA

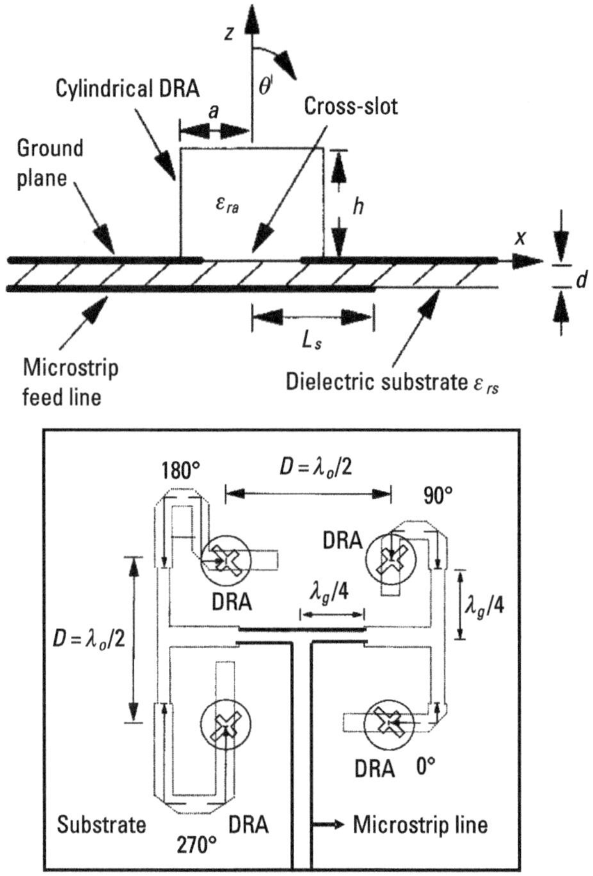

Figure 3.39 The antenna structure proposed in [73].

elements are used to analyze their effect on the CP bandwidth. Further, in [75], a sequentially fed subarray of elliptical DRAs was reported for a wide CP band. Three different types of feeding networks are explored: a parallel feeding network, a series feeding network, and a hybrid ring feeding network for excitation of an antenna array system. A DRA subarray with a hybrid ring feeding network obtained the best results, with a 44% impedance matching bandwidth and a 26% 3-dB AR bandwidth.

Fractal cross-slot-coupled DRA is examined for CP in [76]. By controlling the dimensions of the fractal cross-slot, the resonances of the fractal cross-slot and the dielectric resonator can be merged to obtain a wider AR bandwidth for various wireless applications. The AR bandwidth of the proposed wideband CP DRA array obtained 38.3% (6.06–8.93 GHz) and a 12.17-dBi peak gain.

In [77], the author reports on a circularly polarized substrate-integrated CDRA array for 60-GHz applications. A 2 × 2 substrate-integrated CDRA is designed using $\varepsilon_r = 10.2$ dielectric material. High dielectric material is selected for the design of the upper substrate to excite the DR $HEM_{11\delta}$ mode. Lower-permittivity substrates ($\varepsilon_r = 2.2$) are used to etch the microstrip feed line to excite the DRA. Through two concentric circular arrays of vias and holes, the CDRA is isolated from the dielectric substrate. The substrate-integrated waveguide (SIW) cavity consists of the vias. The inside aperture is a circular patch with a pair of 45° titled slits responsible for generating the CP fields in the DRA. To excite the antenna, a rectangular slot with a length of l_s and a width of w_s is fabricated at the center of the ground plane between the two substrates. A 50-Ω feed line having a width of W_f is printed at the bottom side of the lower substrate. Figure 3.40 shows the configuration of the substrate-integrated CDRA. To improve the 3-dB AR bandwidth, a sequential-rotation method is applied to excite the DRA elements. The DRA array covers the 60-GHz frequency band (57–64 GHz) and 15.9% 3-dB AR bandwidths (55–64.5 GHz). The boresight antenna gain is stable across the operating band, and an 11.43-dBi peak value is obtained.

A circularly polarized wideband rectangular DRA array, excited by microstrip lines, is proposed for a microwave image–sensing application in [78]. Basically, two orthogonal microstrip lines are designed in substrate to excite a single element of a DRA for achieving CP. A four-element DRA linear array is used to obtain higher directive gain with CP for various wireless applications. The antenna is fabricated by Emerson & Cuming Eccostock HIK material, which has a dielectric constant $\varepsilon_r = 20$; the substrate consists of Arlon AD270 dielectric material, with a thickness of 0.79 mm, a loss tangent of 0.0023, and a dielectric constant of $\varepsilon_r = 2.7$. A power divider is designed by the quarter-wavelength transformer to excite the DRA. Figure 3.41 presents the proposed 1 × 4 DRA array configuration. A microstrip feed line extra path length excites a 90° phase delay of the feeding signal of the DRA. The DRA excites two orthogonal

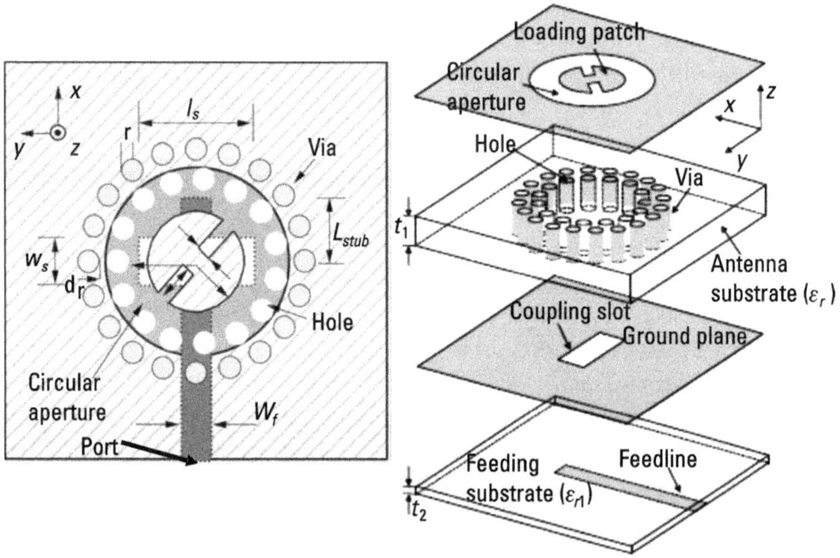

Figure 3.40 The configuration of the substrate-integrated CDRA proposed in [77].

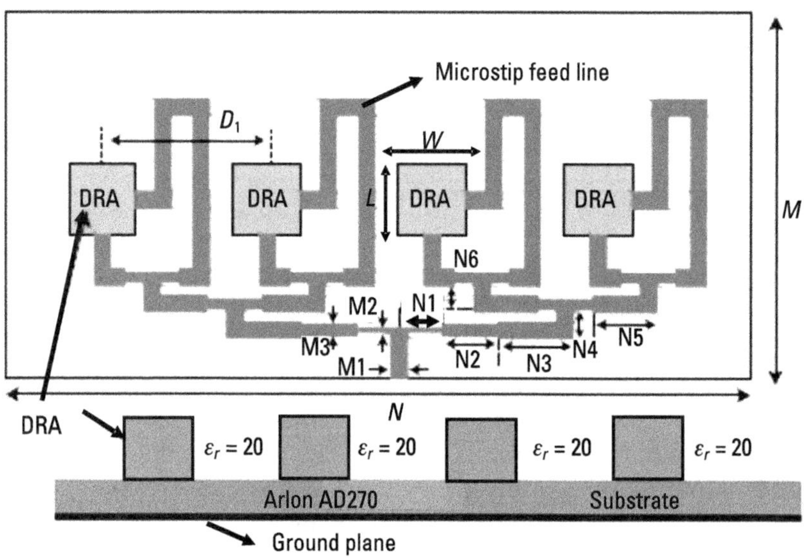

Figure 3.41 The 1 × 4 DRA array configuration proposed in [78].

TE_{111} modes with a 90° phase shift for CP over the operating bandwidth. A measured peak gain of 13.6 dBi and an AR bandwidth of 10% are obtained for the proposed antenna array.

An aperture-coupled modified CDRA array is presented for wideband CP in [79]. The proposed aperture excites the dual-radiating modes $HEM_{11\delta}$ and $TE_{01\delta}$ inside the CDRA. Figure 3.42 shows the fabricated proposed antenna design layout. The antenna prototype is fabricated and experimentally verified. The antenna covers the operating frequency range 4.7–6.4 GHz with a 3-dB AR bandwidth of 23.52% (4.5–5.7 GHz). This antenna is suitable for WLAN (5.2 GHz) and local thermal equilibrium (LTE) band 46 (5.5 GHz) wireless applications.

3.5 CP by Switchable Feeding Techniques

Circular polarized signals are used in satellite, space communication, radar, and navigation systems to allow more flexibility in the inclination of the transmitter and the receiver antennas. For antennas, excited mode field distribution and relative phase differences are generated by switching technologies, micro electromechanical system (MEMS) switches, P-I-N diodes, and photoconductive switches for radio-frequency signals.

Many other geometries with feeding techniques are investigated for achieving wideband with CP in [81–89]. The integration of switchable components with feed line is done to attain the AR bandwidth in [81]. A switchable CDRA resonates at four different radiation modes. The proposed antenna is excited for two linear polarizations, one circular polarized and the second a nonradiating mode for wireless applications. For designing the antenna, a cylindrical-shaped radiator is constructed from zirconium tin titanate, ZrSnTiO, with a high dielectric permittivity of $\varepsilon_r = 37$. Two coaxial feed lines through CDRA are excited, and this feeding network is soldered with a microstrip transmission feed line. In order to design a feeding network, the microstrip feed line is printed

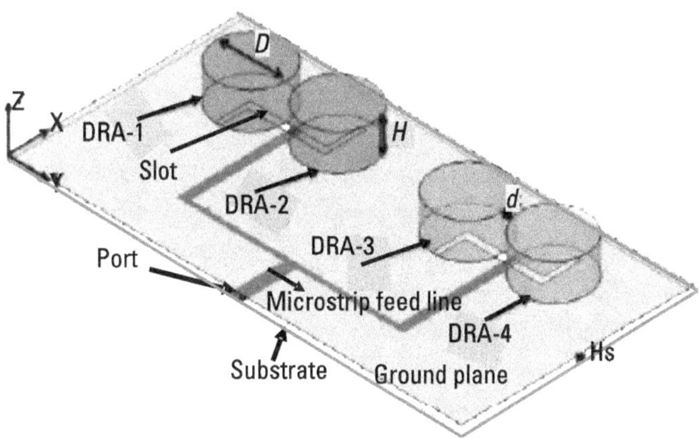

Figure 3.42 The fabricated antenna design layout proposed in [79].

on the dielectric substrate RT/Duroid 5880, which has a dielectric permittivity $\varepsilon_r = 2.2$. To obtain CP, quadrature phase is required in the radiating modes. So a 90° phase shift of the two excitation signals is obtained through the difference in length to the feed lines of the two probes at $\lambda/4$. Figure 3.43 illustrates this antenna configuration and feeding circuit layout. For obtaining a different radiating mode, a switching operation is applied on the feeding network. The switching operation is achieved by two inductors L (47-nH), two resistors R (22-Ω), and two diodes D_1 and D_2.

In the first case of radiating mode, when the power supply V_{cc} is connected to a +2-V DC, the diode D_1, is reverse-biased and operates as an open circuit. Exciting the input signal of the $\lambda/4$ transformer is connected to probe 1 and generates $HE_{11\delta}$ mode in the DRA. If a -2-V DC power supply is applied on V_{cc}, the diode D_2 is forward-biased and operates as a short circuit. Hence, no input signal excites the CDRA (i.e., no radiating mode is generated). Table 3.1 details the switching performance of the feeding network. This antenna obtains a 3% AR bandwidth for CP.

In [82], a probe-fed reconfigurable dual-tripleband Spidron fractal DRA (SFDRA) is considered for CP. $TM_{11\delta}$ and quasi-$TM_{21\delta}$ are generated in the lower frequency band, and the quasi-$TE_{12\delta}$ mode is excited in the upper CP band. The modes of the lower CP band from the triple band can be achieved when it is separated by etching a T-shaped slot from the ground plane. In the circularly polarized band, configurability is achieved by positing a diode D_1 across the width of the T-shaped slot such that the ON-OFF-state of the D_1 radiates the dual or triple band of the CP waves. Figure 3.44 depicts the fabricated prototype of the proposed DRA and feeding network [82]. The antenna is tested, and the results are experimentally verified. The measurement results obtained are based on the ON-OFF state condition of the diode. Impedance bandwidths of 57.60% (2.2–3.98 GHz) at the ON condition and 14.44% (2.12–2.45 GHz) and 40.48% (2.64–3.98 GHz) at the OFF condition are

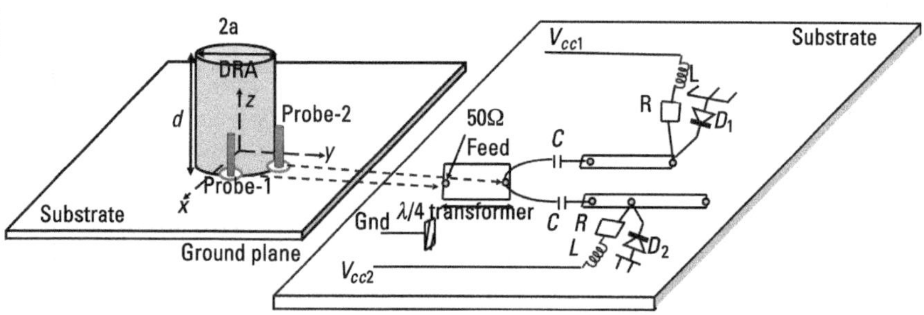

Figure 3.43 The antenna configuration and feeding circuit layout proposed in [81].

Table 3.1
The Switching Performance of the Feeding Network for Different Mode Operations

Mode of operation	V_{cc1}	V_{cc2}	Diode-1	Diode-2
Linear polarization-1	+2V	–2V	Reverse-biased	Forward-biased
Linear polarization-1	–2V	+2V	Forward-biased	Reverse-biased
Circular polarization	+2V	+2V	Reverse-biased	Reverse-biased
Switched off	–2V	–2V	Forward-biased	Forward-biased

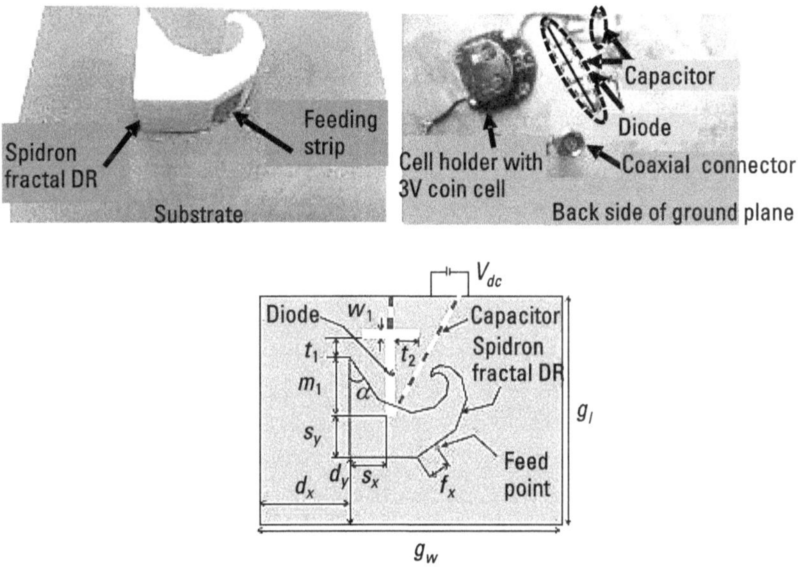

Figure 3.44 Fabricated prototype of the proposed DRA and feeding network [82].

achieved. Similarly, the 3-dB AR bandwidths of 24% (2.31–2.94 GHz) and 6.82% (3.68–3.94 GHz) at the ON condition and 6.34% (2.29–2.44 GHz), 6.65% (2.76–2.95 GHz), and 7.09% (3.67–3.94 GHz) at the OFF condition are obtained.

In [83], an electronically controlled, frequency-agile hybrid DRA is presented for CP. Two L-shaped metallic strips are placed adjacent to a partial ground plane, which can be operated in short or open behavior in four combinations of two P-I-N diodes (D_1 and D_2). Frequency reconfigurability is obtained by switching from both OFF to at least one ON state to yield a single- and dual-band frequency response. The switching ON-OFF state generates a right-handed CP wave at the lower band that can be changed to a left-handed CP wave by switching to the OFF-ON state. Three-dB AR bandwidths of 12.20% (2.54–2.87 GHz) and 10.95% (2.59–2.89 GHz) are obtained respectively when D1 and D2 are turned ON alternatively. Figure 3.45 depicts the geometry of this proposed antenna and its fabricated prototype.

The authors in [84] report on CP in DRA through dual probes with a tunable feeding network. $HEM_{11\delta}$ modes are excited by a dual-probe feeding arrangement in a DRA. Polarization reconfigurability characteristics can be achieved by a tunable feed network, which consists of a Wilkinson equal-power divider, parasitic elements used for phase shifters, and P-I-N diodes. By controlling the four pairs of P-I-N diodes ON-OFF state to active corresponding phase shifters. This phase shifter generates equal amplitude and 0° or 90° phase differences for CP. The proposed DRA achieves a more than 30% impedance bandwidth and a 3-dB AR bandwidth at the CP state and a 6.3% and 9.5% impedance bandwidth for the LP state. Stable directional patterns and gains are achieved for all ON-OFF states. These antenna performances are suitable for polarization diversity scenarios in communication systems. Figure 3.46 shows the proposed antenna geometry and tunable feeding network proposed.

A reconfigurable DRA is investigated for CP in [85]. This antenna consists of a stacked microstrip patch antenna (MPA) with a stacked DRA. With the techniques of polarization reconfigurability, the antenna achieved RHCP, LHCP, and LP. Other new antenna geometry has been explored by researchers for achieving CP. A low-profile reconfigurable water spiral antenna is reported for CP in [86]. In this case, an Archimedean spiral antenna with two water arms is excited by a parallel strip line. A glass container with two water channels is placed above the reflecting plane, the upper one for LHCP and the lower one for RHCP. Polarization of the antenna can be controlled by the flow of water from the water channels. Figure 3.47 shows the configuration of this Archimedean spiral. The antenna covers a 40% impedance bandwidth and an AR bandwidth potentially useful for Global Positioning System (GPS) and BeiDou Naviation Satellite System (BDS) application. Similarly in [87], an asymmetric liquid flow-control polarization reconfigurable CDRA for wireless communications is investigated. The CDRA is fabricated by ceramic material and two pairs of asymmetrical holes drilled on the 45° and 135° axis of the CDRA. By

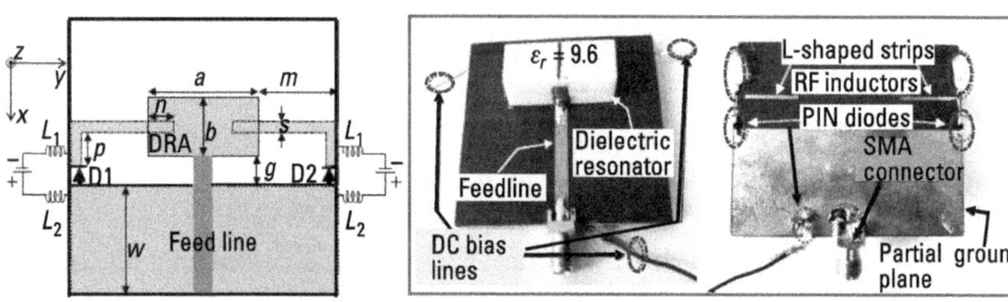

Figure 3.45 Switchable CP DRA proposed in [83]: (a) the geometry of the antenna and (b) the fabricated prototype.

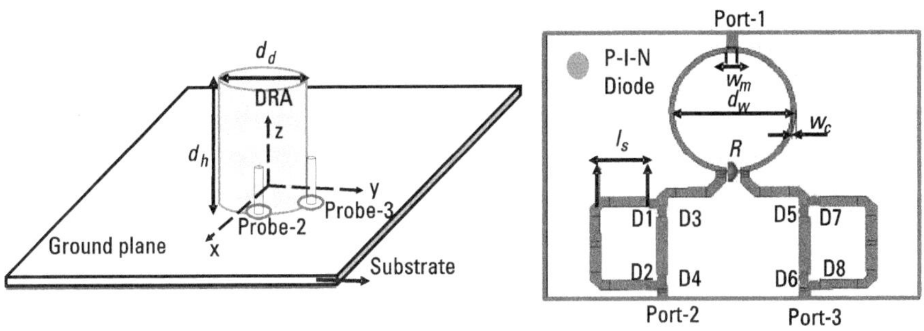

Figure 3.46 The antenna geometry and tunable feeding network proposed in [84].

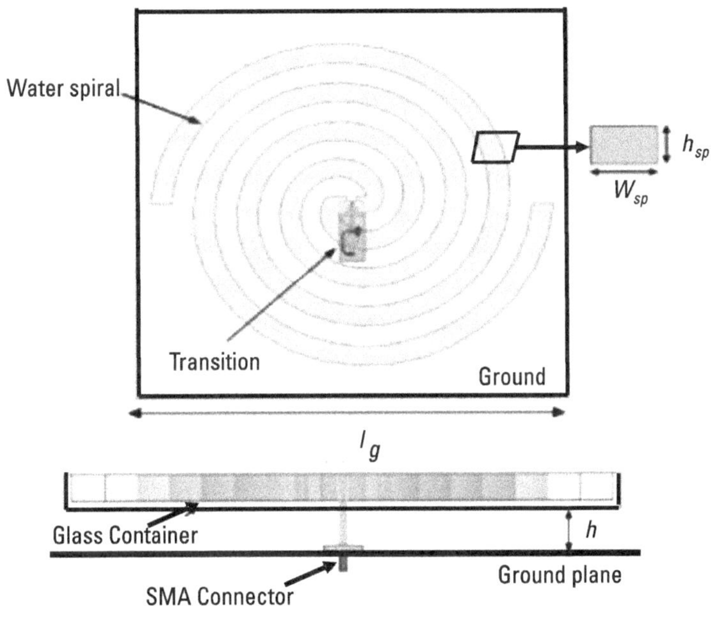

Figure 3.47 Configuration of Archimedean spiral [86].

alternating liquid injection and extraction in the two pairs of holes, different cases of polarization are obtained.

In addition, [88] investigates an electronically steerable parasitic array radiator of a DRA, and [89] explores a wideband CP beam steering DRA using a gravitational ball lens. The radiation patterns of the base antenna can be controlled by the gravity of the ball lenses and the base antenna is mounted on the wall or ceiling. Figure 3.48 illustrates this antenna concept geometry, which achieved a 42% impedance bandwidth and a 39% AR bandwidth.

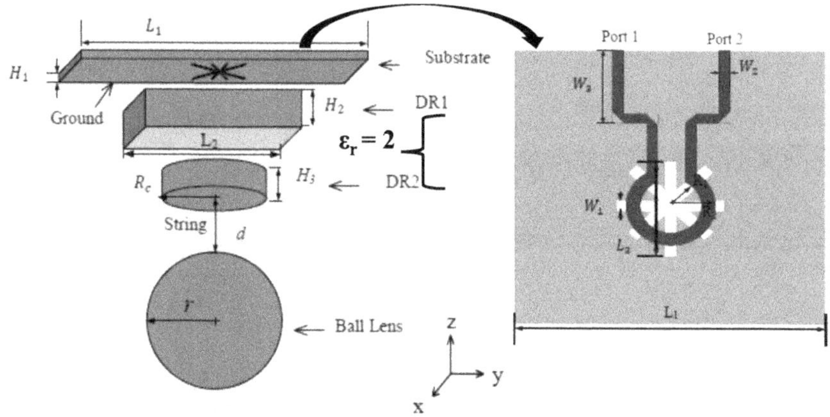

Figure 3.48 The antenna geometry proposed in [89].

3.6 Conclusion

This chapter discussed single-feeding techniques, multiple-feeding excitation, modified antenna geometries, the arrays of various antenna geometries, and the switchable feeding techniques used to achieve CP. In addition, the chaper explained some basic CP geometry and feeding techniques. Moreover, the chapter included a detailed analysis of some novel circularly polarized antennas like the water Archimedean spiral, reconfigurable DRA, substrate-integrated DRA, and sequential-fed DRA. In addition, the chapter covered the circular polarized antenna design process and results.

References

[1] Balanis, C. A., *Antenna Theory: Analysis, and Design,* Hoboken, NJ: Wiley, 1997.

[2] Haneishi, M., and Y. Suzuki, "Circular Polarisation and Bandwidth" (Chapter 4) in *Handbook of Microstrip Antennas,* J. R. James and P. S. Hall (eds.), London, England: Peregrin Press, 1989.

[3] Oliver, M. B., et al., "Circularly Polarised Rectangular Dielectric Resonator Antenna," *IET Electronics Letters,* Vol. 31, No. 6, March 16, 1995, pp. 418–419.

[4] Esselle, K. P., "Circularly Polarised Higher-Order Rectangular Dielectric-Resonator Antenna," *Electronics Letters,* Vol. 32, No. 3, Feb. 1, 1996, pp. 150–151.

[5] Chowdhury, R., et al., "Analysis of a Wideband Circularly Polarized Cylindrical Dielectric Resonator Antenna with Broadside Radiation Coupled with Simple Microstrip Feeding," *IEEE Access,* Vol. 5, 2017, pp. 19478–19485.

[6] Almpanis, G., C. Fumeaux, and R. Vahldieck, "Offset Cross-Slot-Coupled Dielectric Resonator Antenna for Circular Polarization," *IEEE Microwave and Wireless Components Letters*, Vol. 16, No. 8, Aug. 2006, pp. 461–463.

[7] Li, B., and K. W. Leung, "On the Circularly Polarized Hemi-Ellipsoidal Dielectric Resonator Antenna," *Microwave and Optical Technology Letters*, Vol. 48, No. 9, pp. 1763–1766.

[8] Kumar, R., and R. K. Chaudhary, "Investigation of Higher Order Modes Excitation Through F–Shaped Slot in Rectangular Dielectric Resonator Antenna for Wideband Circular Polarization with Broadside Radiation Characteristics," *Int J RF Microw Comput Aided Eng.*, Vol. 28, No 6, 2018.

[9] Pan, Y., and K. W. Leung, "Wideband Circularly Polarized Trapezoidal Dielectric Resonator Antenna," *IEEE Antennas and Wireless Propagation Letters*, Vol. 9, 2010, pp. 588–591.

[10] Li, B., K. K. So, and K. W. Leung, "A Circularly Polarized Dielectric Resonator Antenna Excited by an Asymmetrical U–Slot with a Backing Cavity," *IEEE Antennas and Wireless Propagation Letters*, Vol. 2, 2003, pp. 133–135.

[11] Zou, M., et al., "A Wideband Circularly Polarized Rectangular Dielectric Resonator Antenna Excited by a Lumped Resistively Loaded Monofilar-Spiral-Slot," *IEEE Antennas and Wireless Propagation Letters*, Vol. 12, 2013, pp. 1646–1649.

[12] Luk, K. M., and K. W. Leung, *Dielectric Resonator Antennas*, London, U.K.: Research Studies Press, 2003.

[13] Zou, M., J. Pan, and Z. Nie, "A Wideband Circularly Polarized Rectangular Dielectric Resonator Antenna Excited by an Archimedean Spiral Slot," *IEEE Antennas and Wireless Propagation Letters*, Vol. 14, 2015, pp. 446–449.

[14] Leung, K. W., and K. K. So, "Frequency–Tunable Designs of the Linearly and Circularly Polarized Dielectric Resonator Antennas Using a Parasitic Slot," *IEEE Transactions on Antennas and Propagation*, Vol. 53, No. 1, Jan. 2005, pp. 572–578.

[15] Wang, L., et al., "A New Broadband Circularly Polarized Dielectric Resonator Antenna: Using Parasitic Elements," *AEU–International Journal of Electronics and Communications*, Vol. 128, 2021.

[16] Leung, K. W., and H. K. Ng, "Theory and Experiment of Circularly Polarized Dielectric Resonator Antenna with A Parasitic Patch," *IEEE Transactions on Antennas and Propagation*, Vol. 51, No. 3, March 2003, pp. 405–412.

[17] Sulaiman, M. I., and S. K. Khamas, "A Singly Fed Wideband Circularly Polarized Dielectric Resonator Antenna Using Concentric Open Half–Loops," *IEEE Antennas and Wireless Propagation Letters*, Vol. 10, 2011, pp. 1305–1308.

[18] Khamas, S. K., "Circularly Polarized Dielectric Resonator Antenna Excited by a Conformal Wire," *IEEE Antennas and Wireless Propagation Letters*, Vol. 7, 2008, pp. 240–242.

[19] Motevasselian, A., A. Ellgardt, and B. L. G. Jonsson, "A Circularly Polarized Cylindrical Dielectric Resonator Antenna Using a Helical Exciter," *IEEE Transactions on Antennas and Propagation*, Vol. 61, No. 3, March 2013, pp. 1439–1443.

[20] Ding, Y., K. W. Leung, and K. M. Luk, "Compact Circularly Polarized Dualband Zonal–Slot/DRA Hybrid Antenna Without External Ground Plane," *IEEE Transactions on Antennas and Propagation*, Vol. 59, No. 6, June 2011, pp. 2404–2409.

[21] Guo, L., and K. W. Leung, "Compact Unilateral Circularly Polarized Dielectric Resonator Antenna," *IEEE Transactions on Antennas and Propagation,* Vol. 66, No. 2, Feb. 2018, pp. 668–674.

[22] Chauhan, M., A. Rajput, and B. Mukherjee, "Wideband Circularly Polarized Low Profile Dielectric Resonator Antenna With Meta Superstrate for High Gain," *AEU–International Journal of Electronics and Communications,* Volume 128, 2021.

[23] Gangwar, D., et al., "Frequency Selective Surface as Superstrate on Wideband Dielectric Resonator Antenna for Circular Polarization and Gain Enhancement," *Wireless Pers Commun,* 2017.

[24] Petosa, A., and A. Ittipiboon, "Dielectric Resonator Antennas: A Historical Review and the Current State of the Art," *IEEE Antennas and Propagation Magazine,* Vol. 52, No. 5, Oct. 2010, pp. 91–116.

[25] Ittipiboon, A., A. Petosa, and M. Cuhaci, "Effects of Interfeed Coupling on the Axial Ratio Performance of Circular Polarized Microstrip Antennas," *1998 Symposium on Antenna Technology and Applied Electromagnetics, Ottawa, Canada,* 1998, pp. 209–212.

[26] Mongia, R. K., et al., "Circularly Polarised Dielectric Resonator Antenna," *Electronics Letters,* Vol. 30, No. 17, Aug. 18, 1994, pp. 1361–1362.

[27] Drossos, G., Z. Wu, and L. E. Davis, "Circular Polarised Cylindrical Dielectric Resonator Antenna," *Electronics Letters,* Vol. 32, No. 4, Feb. 15, 1996, pp. 281–283.

[28] Buerkle, A., and K. Sarabandi, "A Wideband, Circularly Polarized, Magnetodielectric Resonator Antenna," *IEEE Transactions on Antennas and Propagation,* Vol. 53, No. 11, Nov. 2005, pp. 3436–3442.

[29] Huang, C.-Y., Chiou, T.-W., and Wong, K.-L., "Dual–Polarized Dielectric Resonator Antennas," *Microwave and Optical Technology Letters,* Vol. 31, No. 3, 2001, pp. 222–223.

[30] Gray, D., and T. Watanabe, "Three Orthogonal Polarisation DRA–Monopole Ensemble," *Electronics Letters,* Vol. 39, No. 10, May 15, 2003, pp. 766–767.

[31] Li, B., C. Hao, and X. Sheng, "A Dual-Mode Quadrature-Fed Wideband Circularly Polarized Dielectric Resonator Antenna," *IEEE Antennas and Wireless Propagation Letters,* Vol. 8, 2009, pp. 1036–1038.

[32] Zhao, Z., et al., "Wideband Dual-Feed, Dual-Sense Circularly Polarized Dielectric Resonator Antenna," *IEEE Transactions on Antennas and Propagation,* Vol. 68, No. 12, Dec. 2020, pp. 7785–7793.

[33] Leung, K. W., et al., "Circular-Polarised Dielectric Resonator Antenna Excited by Dual Conformal Strips," *Electronics Letters,* Vol. 36, No. 6, March 16, 2000, pp. 484–486.

[34] Khoo, K., Y. Guo, and L. C. Ong, "Wideband Circularly Polarized Dielectric Resonator Antenna," *IEEE Transactions on Antennas and Propagation,* Vol. 55, No. 7, July 2007, pp. 1929–1932.

[35] Pan, Y., K. W. Leung, and E. H. Lim, "Compact Wideband Circularly Polarized Rectangular Dielectric Resonator Antenna with Dual Underlaid Hybrid Couplers," *Microwave and Optical Technology Letters,* Vol. 52, No. 12, 2010, pp. 2789–2791.

[36] Lim, E. H., K. W. Leung, and X. S. Fang, "The Compact Circularly Polarized Hollow Rectangular Dielectric Resonator Antenna with an Underlaid Quadrature Coupler," *IEEE Transactions on Antennas and Propagation,* Vol. 59, No. 1, Jan. 2011, pp. 288–293.

[37] Xiang, B. J., et al., "Wideband Circularly Polarized Dielectric Resonator Antenna with Bandpass Filtering and Wide Harmonics Suppression Response," *IEEE Transactions on Antennas and Propagation,* Vol. 65, No. 4, April 2017, pp. 2096–2101.

[38] Massie, G., et al., "A New Wideband Circularly Polarized Hybrid Dielectric Resonator Antenna," *IEEE Antennas and Wireless Propagation Letters,* Vol. 9, 2010, pp. 347–350.

[39] Liu, J., and S.-S. Zhong, "Broadband Circularly Polarised Dielectric Resonator Antenna Fed by Wideband Switched Line Coupler," *Electronics Letters,* Vol. 5, No. 10, 2014, pp. 725–726.

[40] Lu, L., et al., "A Novel Low-Profile Dual Circularly Polarized Dielectric Resonator Antenna," *IEEE Transactions on Antennas and Propagation,* Vol. 64, No. 9, Sept. 2016, pp. 4078–4083.

[41] Tam, M. T. K., and R. D. Murch, "Circularly Polarized Circular Sector Dielectric Resonator Antenna," *IEEE Transactions on Antennas and Propagation,* Vol. 48, No. 1, Jan. 2000, pp. 126–128.

[42] Lee, J. M., et al., "Circularly Polarized Semi-Eccentric Annular Dielectric Resonator Antenna for X–Band Applications," *IEEE Antennas and Wireless Propagation Letters,* Vol. 14, 2015, pp. 1810–1813.

[43] Chowdhury, R., and R. K. Chaudhary, "Investigation on Different Forms of Circular Sectored–Dielectric Resonator Antenna for Improvement in Circular Polarization Performance," *IEEE Transactions on Antennas and Propagation,* Vol. 66, No. 10, Oct. 2018, pp. 5596–5601.

[44] Patel, P., B. Mukherjee, and J. Mukherjee, "Wideband Circularly Polarized Rectangular Dielectric Resonator Antennas Using Square–Shaped Slots," *IEEE Antennas and Wireless Propagation Letters,* Vol. 15, 2016, pp. 1309–1312.

[45] Altaf, A., et al., "Circularly Polarized Spidron Fractal Dielectric Resonator Antenna," *IEEE Antennas and Wireless Propagation Letters,* Vol. 14, 2015, pp. 1806–1809.

[46] Varshney, G., et al., "Inverted–Sigmoid Shaped Multiband Dielectric Resonator Antenna with Dual–Band Circular Polarization," *IEEE Transactions on Antennas and Propagation,* Vol. 66, No. 4, April 2018, pp. 2067–2072.

[47] Chauhan, M., A. K. Pandey, and B. Mukherjee, "A Novel Cylindrical Dielectric Resonator Antenna Based on Fibonacci Series Approach," *Microwave and Optical Technology Letters,* Wiley, Vol. 61, Issue 10, 2019, pp. 2268–2274.

[48] Fakhte, S., H. Oraizi, and R. Karimian, "A Novel Low-Cost Circularly Polarized Rotated Stacked Dielectric Resonator Antenna," *IEEE Antennas and Wireless Propagation Letters,* Vol. 13, 2014, pp. 722–725.

[49] Sharma, A., and R. K. Gangwar, "Circularly polarised hybrid Z–shaped cylindrical Dielectric Resonator Antenna for Multiband Applications," *IET Microwaves, Antennas and Propagation,* Vol. 10, No. 12, 2016, pp. 1259–1267.

[50] Fakhte, S., et al., "A New Wideband Circularly Polarized Stair-Shaped Dielectric Resonator Antenna," *IEEE Transactions on Antennas and Propagation,* Vol. 63, No. 4, April 2015, pp. 1828–1832.

[51] Wang, K. X., and H. Wong, "A Circularly Polarized Antenna by Using Rotated–Stair Dielectric Resonator," *IEEE Antennas and Wireless Propagation Letters,* Vol. 14, 2015, pp. 787–790.

[52] Chauhan, M., and B. Mukherjee, "Wideband Circular Polarized Cylindrical Segmented Dielectric Resonator Antenna for ISM Band Applications," *Int J RF Microw Comput Aided Eng.,* 2020, Vol. 30, No. 4, 1–10.

[53] Zou, M., and J. Pan, "Wide Dual-Band Circularly Polarized Stacked Rectangular Dielectric Resonator Antenna," *IEEE Antennas and Wireless Propagation Letters,* Vol. 15, 2016, pp. 1140–1143.

[54] Pan, Y. M., and K. W. Leung, "Wideband Omnidirectional Circularly Polarized Dielectric Resonator Antenna with Parasitic Strips," *IEEE Transactions on Antennas and Propagation,* Vol. 60, No. 6, June 2012, pp. 2992–2997.

[55] Fang, X., K. W. Leung, and E. H. Lim, "Singly Fed Dual-Band Circularly Polarized Dielectric Resonator Antenna," *IEEE Antennas and Wireless Propagation Letters,* Vol. 13, 2014, pp. 995–998.

[56] Khalily, M., et al., "Omnidirectional Circularly Polarized Dielectric Resonator Antenna for 5.2-GHz WLAN Applications," *IEEE Antennas and Wireless Propagation Letters,* Vol. 13, 2014, pp. 443–446.

[57] Varshney, G., et al., "Wide Band Circularly Polarized Dielectric Resonator Antenna with Stair–Shaped Slot Excitation," *IEEE Transactions on Antennas and Propagation,* Vol. 65, No. 3, March 2017, pp. 1380–1383.

[58] Sharma, A., and R. K. Gangwar, "Circularly Polarised Hybrid Z–Shaped Cylindrical Dielectric Resonator Antenna for Multiband Applications," *IET Microwaves, Antennas & Propagation,* Vol. 10, No. 12, 2016, pp. 1259–1267.

[59] Perron, A., T. A. Denidni, and A. R. Sebak, "Circularly Polarized Microstrip/Elliptical Dielectric Ring Resonator Antenna for Millimeter-Wave Applications," *IEEE Antennas and Wireless Propagation Letters,* Vol. 9, 2010, pp. 783–786.

[60] Khalily, M., M. K. A. Rahim, and A. A. Kishk, "Planar Wideband Circularly Polarized Antenna Design with Rectangular Ring Dielectric Resonator and Parasitic Printed Loops," *IEEE Antennas and Wireless Propagation Letters,* Vol. 11, 2012, pp. 905–908.

[61] Sharma, A., et al., "Quad-Band Quad-Sense Circularly Polarized Dielectric Resonator Antenna for GPS/CNSS/WLAN/WiMAX Applications," *IEEE Antennas and Wireless Propagation Letters,* Vol. 19, No. 3, March 2020, pp. 403–407.

[62] Dash, U. A., and S. Sahu, "Wideband Circularly Polarized Sectored Conical Dielectric Resonator Antenna Excited by S–Shaped Slot," *Microwave and Optical Technology Letters,* Vol. 60, No. 2, 2018, pp. 310–318.

[63] Kumar, R., and R. K. Chaudhary, "A Wideband Circularly Polarized Cubic Dielectric Resonator Antenna Excited with Modified Microstrip Feed," *IEEE Antennas and Wireless Propagation Letters,* Vol. 15, 2016, pp. 1285–1288.

[64] Li, W. W., and K. W. Leung, "Omnidirectional Circularly Polarized Dielectric Resonator Antenna with Top–Loaded Alford Loop for Pattern Diversity Design," *IEEE Transactions on Antennas and Propagation,* Vol. 61, No. 8, Aug. 2013, pp. 4246–4256.

[65] Pan, Y. M., S. Y. Zheng, and W. Li, "Dual-Band and Dual-Sense Omnidirectional Circularly Polarized Antenna," *IEEE Antennas and Wireless Propagation Letters,* Vol. 13, 2014, pp. 706–709.

[66] Dash, S. K. K., T. Khan, and B. K. Kanaujia, "Circularly Polarized Dual Facet Spiral Fed Compact Triangular Dielectric Resonator Antenna for Sensing Applications," *IEEE Sensors Letters,* Vol. 2, No. 1, March 2018, pp. 1–4.

[67] Kakade, A. B., and M. S. Kumbhar, "Wideband Circularly Polarized Conformal Strip Fed Three Layer Hemispherical Dielectric Resonator Antenna with Parasitic Patch," *Microwave and Optical Technology Letters,* Vol. 56, No. 1, 2013, pp. 72–77.

[68] Singhwal, S. S., et al., "Circularly Polarized V-Shaped Dielectric Resonator Antenna," *Int J RF Microw Comput Aided Eng.,* 2019.

[69] Petosa, A., *Dielectric Resonator Antenna Handbook,* Norwood, MA: Artech House, 2007.

[70] Haneishi, M., and H. Takazawa, "Broadband Circularly Polarised Planar Array Composed of a Pair of Dielectric Resonator Antennas," *Electronics Letters,* Vol. 21, No. 10, May 9, 1985, pp. 437–438.

[71] Petosa, A., A. Ittipiboon, and M. Cuhaci, "Array of Circular-Polarised Cross-Dielectric Resonator Antennas," *Electronics Letters,* Vol. 32, No. 19, 12 Sept. 1996, pp. 1742–1743.

[72] Li, B., and K. W. Leung, "On the Circularly Polarized Hemi-ellipsoidal Dielectric Resonator Antenna," *Microwave and Optical Technology Letters,* Vol. 48, No. 9, 2006, pp. 1763–1766.

[73] Laisne, A., R. Gillard, and G. Piton, "Circularly Polarised Dielectric Resonator Antenna with Metallic Strip," *Electronics Letters,* Vol. 38, No. 3, Jan. 31, 2002, pp. 106–107.

[74] Kishk, A. A., "Performance of Planar Four–Element Array of Single–Fed Circularly Polarized Dielectric Resonator Antenna," *Microwave and Optical Technology Letters,* Vol. 38, No. 5, 2003, pp. 381–384.

[75] Yang, S. S., et al., "Study on Sequential Feeding Networks for Subarrays of Circularly Polarized Elliptical Dielectric Resonator Antenna," *IEEE Transactions on Antennas and Propagation,* Vol. 55, No. 2, Feb. 2007, pp. 321–333.

[76] Lin, J.-H., et al., "Circularly Polarized Dielectric Resonator Antenna Arrays with Fractal Cross-Slot-Coupled DRA Elements," *International Journal of Antennas and Propagation,* Hindwai, Vol, 2017, https://doi.org/10.1155/2017/8160768.

[77] Sun, Y., and K. W. Leung, "Circularly Polarized Substrate-Integrated Cylindrical Dielectric Resonator Antenna Array for 60-GHz Applications," *IEEE Antennas and Wireless Propagation Letters,* Vol. 17, No. 8, Aug. 2018, pp. 1401–1405.

[78] Rana, B., and S. K. Parui, "Microstrip Line Fed Wideband Circularly Polarized Dielectric Resonator Antenna Array for Microwave Image Sensing," *IEEE Sensors Letters,* Vol. 1, No. 3, June 2017, pp. 1–4.

[79] Gupta, S., et al., "Wideband Circularly Polarized Dielectric Resonator Antenna Array with Polarization Diversity," *IEEE Access,* Vol. 7, 2019, pp. 49069–49076.

[80] Shi, W., et al., "Design of a Broadband Circularly Polarized Dielectric Resonator Antenna Array in C-Band," *2017 Sixth Asia–Pacific Conference on Antennas and Propagation (APCAP),* Xi'an, 2017, pp. 1–3.

[81] Drossos, G., Z. Wu, and L. E. Davis, "Switchable Cylindrical Dielectric Resonator Antenna," *Electronics Letters,* Vol. 32, No. 10, May 9, 1996, pp. 862–864.

[82] Altaf, A., et al., "Reconfigurable Dual-Triple-Band Circularly Polarized Dielectric Resonator Antenna," *IEEE Antennas and Wireless Propagation Letters,* Vol. 19, No. 3, March 2020, pp. 443–447.

[83] Altaf, A., and M. Seo, "An Electronically Switched Frequency-Agile Hybrid Dielectric Resonator Antenna with a Fixed Reconfigurable Circularly Polarized Band," *IEEE Access,* Vol. 8, 2020, pp. 143509–143518.

[84] Liu, B., et al., "Polarization-Reconfigurable Cylindrical Dielectric Resonator Antenna Excited by Dual Probe with Tunable Feed Network," *IEEE Access,* Vol. 7, 2019, pp. 60111–60119.

[85] Sun, W., et al., "A Circularly Polarized Dielectric Resonator Antenna and Its Reconfigurable Design," *IEEE Antennas and Wireless Propagation Letters,* Vol. 19, No. 7, July 2020, pp. 1088–1092.

[86] Hu, Z., et al., "Broadband Polarization-Reconfigurable Water Spiral Antenna of Low Profile," *IEEE Antennas and Wireless Propagation Letters,* Vol. 16, 2017, pp. 1377–1380.

[87] Malhat H.-A., and S. H. ZainudDeen, "Asymmetric Liquid Flow Control Polarization Reconfigurable Cylindrical Dielectric Resonator Antenna for Wireless Communications," *Int J RF Microw Comput Aided Eng.,* 2020.

[88] Movahedinia, R., et al., "X-Band Circularly Polarized Electronically Steerable Parasitic Array Radiator of DRA," *IEEE Transactions on Antennas and Propagation,* Vol. 66, No. 2, Feb. 2018, pp. 721–728.

[89] Chen, Z., et al., "A Wideband Circular-Polarized Beam Steering Dielectric Resonator Antenna Using Gravitational Ball Lens," *IEEE Transactions on Antennas and Propagation,* Vol. 69, No. 5, 2020, pp. 2963–2968.

4

DRA Gain-Enhancement Techniques

In the era of modern wireless technology, high-efficiency, high-gain, directional antennas are required for long-distance communication. High-gain antennas, which enhance the performance of radiation in a particular direction and reduce unwanted interfering signals, are commonly preferred for space and satellite communication. High-gain transmitter antennas allow more transmitted power signals to the receiver, thereby increasing signal strength. In addition, they facilitate the line-of-sight (LOS) communication at the high-frequency range, capture more signal strength, and improve the signal-to-noise ratio (SNR).

- *Directivity and gain:* Directivity is a quantitative measure of an antenna's ability to concentrate maximum energy in a particular direction. It is defined as the ratio of the maximal radiation intensity of the reference antenna to the average radiation intensity of the reference antenna [1] and shown as follows:

$$\text{Directivity} = \frac{\text{maximum radiation intensity in particular direction}}{\text{average radaition intensity}} \tag{4.1}$$

$$\text{Directivity} = \frac{\phi(\theta,\varphi)_{max}}{\phi_{av}} \tag{4.2}$$

Here, $\varphi(\theta, \varphi)_{\max}$ depicts the power per unit of the solid angle, denoting the maximum radiation intensity, and φ_{av} represents the average radiation intensity. Also, the directivity can be defined in terms of power as

$$\text{Directivity} = \frac{\text{power radiated by test antenna}}{\text{power radiated by isotropic antenna}} \quad (4.3)$$

The average radiation intensity is defined as the total radiated power divided by a 4π steradian or the solid angle. Hence, the directivity can also be represented as

$$\text{Directivity} = \frac{\phi(\theta,\varphi)_{max}}{W_r/4\pi} \; or \; \frac{4\pi(\text{maximum radiation intensity})}{\text{total radiated power}} \quad (4.4)$$

Gain is defined as the antenna energy radiated in a particular direction with respect to a standard antenna. A standard antenna includes all the losses, including ohmic losses and dielectric losses. Gain takes into account the antenna efficiency, which is shown as

$$\text{Gain} = \frac{\text{radiation intensity}}{\text{avg radiated power}} = \frac{\phi(\theta,\varphi)}{W_r/4\pi} \quad (4.5)$$

$$\text{Gain} = \eta \times \text{Directivity} \quad (4.6)$$

where η is the radiation efficiency factor. For an isotropic antenna, the gain and directivity are the same since $\eta = 1$.

- *Effective aperture area:* The effective aperture area defines how much EM signal is captured effectively compared to the actual physical aperture. The gain of an antenna is easily calculated by knowing the aperture area of the antenna, as

$$A_{eff} = \lambda^2/4\pi * Gain \quad (4.7)$$

where λ is the wavelength, and A_{eff} is the effective aperture area. The effective aperture of the antenna is generally less than the physical aperture. The larger the effective aperture area, the higher is the gain of the antenna.

- *Perturbation theory and radiation boundary:* If small volume v is removed from the basic geometry of antenna, volume perturbation occurs in antenna. Hence, the resonant frequency is shifted for the proposed DRA, and simultaneously a larger radiation boundary is created for radiating from the antenna. The shifted frequency is calculated by the cavity perturbation theory [2,3].

  When a small volume (Δv) sample is inserted in or removed from a cavity (Figure 4.1), the E field E_0 and M field H_0 (distribution) in the cavity is changed, affecting the radiation of cavity. From Figure 4.1, it can be clearly observed that the E field E_0 and magnetic field H_0 in the unperturbed state and the fields in the inserted/perturbed of the cavity is E and H. If the inserted or removed sample is lossless, then the variation of the resonant frequency is calculated by [2,3]

$$\frac{f_r - f_0}{f_r} = \frac{\iiint \left(\Delta\varepsilon E.E_0^* + \Delta\mu H.H_0^* \right).d\tau}{\iiint \left(\varepsilon E.E_0^* + \mu H.H_0^* \right).d\tau} \tag{4.8}$$

  where f_r is the resonant frequency, ε and μ are the permittivity and permeability of the medium in the unperturbed cavity, $d\tau$ is the elementary volume, and $\Delta\varepsilon$ and $\Delta\mu$ are the modified permittivity and permeability due to the insertion or removal of the sample in the cavity.

4.1 Gain Enhancement by Loading a Metamaterial or Frequency-Selective Surface on a DRA

Various DRAs with metamaterial superstrate loading configurations are investigated to attain high gain [4–7]. In [4], a dual-segment RDRA with metamaterial loading is proposed for high gain and wide bandwidth. A metamaterial superstrate is designed by the array of the unit cell circular-shaped split-ring resonator with conducting strip. This metamaterial unit cell arrangement en-

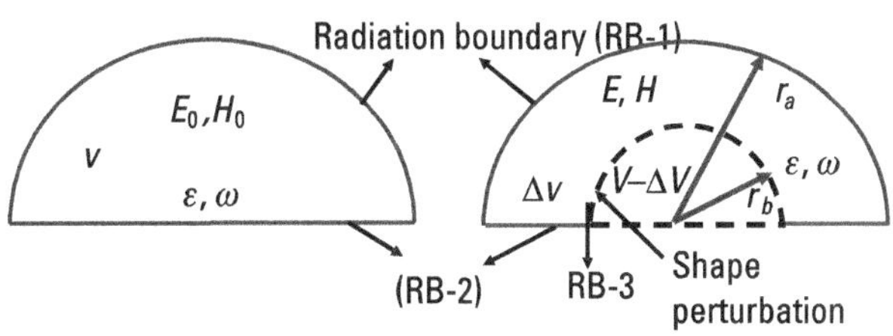

Figure 4.1 E and M field distribution on a radiation boundary: (a) unperturbed cavity and (b) perturbed cavity.

hances the antenna gain by as much as 31% and simultaneously improves the bandwidth to 22%. The antenna is fabricated by stacking two dielectric materials (Teflon $\varepsilon_r = 2.1$ and alumina $\varepsilon_r = 9.8$). For excitation of the antenna, an aperture-coupling technique is used. Figure 4.2(a) shows the geometry of the antenna, and Figure 4.2(b) shows the unit cell of the metamaterial.

A double-negative material property (negative permeability and permittivity) is obtained by an artificially realized metamaterial structure. The split-spring resonator (SRR) and conducting strips are incorporated on opposite sides of the substrate. The SRRs' structure behaves magnetically, and the conducting strips behave electrically to EM fields. To obtain the best antenna results, the height between the substrate and the superstrate metamaterial is optimized. The measured peak gain of the antenna is achieved with and without metamaterial loading of 9.8 dBi (at 5.3 GHz) and 7.51 dBi (at 5.7 GHz). Similarly, in [5], a DRA gain is improved by the loading of a 5 × 5 unit-cell SRR as the superstrate. In the near-field region, the coupling of SSRs with an RDRA induces opposite-direction currents in split rings, which effectively cancel the transverse fields. Figure 4.3(a) shows the current distribution in a unit-cell SRR. A maximum gain of 8 dBi is obtained at a 4-GHz resonant frequency, and overall a 1.5-dBi gain enhancement is achieved over the entire performance bandwidth. The proposed antenna is constructed with Rogers R03010 dielctric material (ε_r =10.2), and the fabricated SRR's array is placed at a certain distance above the RDRA so as to achieve significant results. Figure 4.3(b) shows the fabricated antenna prototype.

In [6], a wideband differential Fabry-Pérot cavity antenna is designed to improve the gain of an antenna [6]. A partially reflective surface (PRS) is used as superstrate to achieve a boresight gain of up to 9.95 dBi. In the proposed antenna, a differential feeding technique offers high noise immunity compared

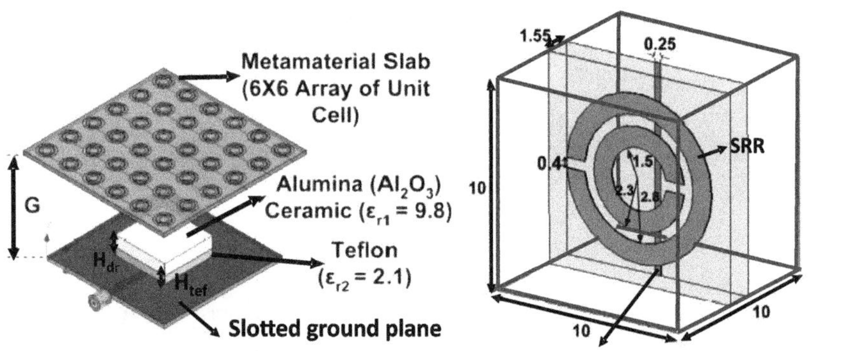

(a) (b)

Figure 4.2 DRA with superstrate: (a) geometry of the antenna and (b) unit cell of the metamaterial [4].

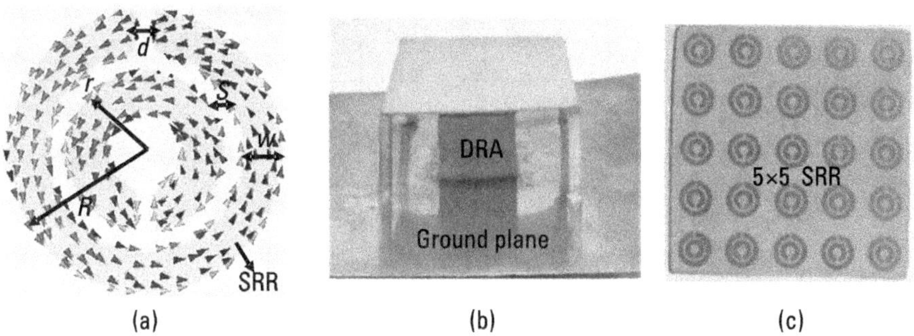

Figure 4.3 RDRA with an SRR: (a) current distribution in a unit-cell SRR and (b) the fabricated antenna prototype [5].

to a single feeding network. Placing the PRS at an optimum distance from the DRA offers the best antenna performance for wireless communication. A PRS is designed by repetitive arrangement of a unit cell. The single-side printed PRS is constructed with 0.77-mm-thick GML 1000, which has a dielectric permittivity $\varepsilon_r = 3.2$. The PRS magnitude of reflection coefficient is best chosen to enhance the gain of the antenna. The 3-dB bandwidth performance of the antenna is shown by

$$G = \frac{1+|S_{11}|}{1-|S_{11}|} \tag{4.9}$$

$$B.W._{(3dB)} = \frac{\lambda}{2\pi h_{PRs}} * \frac{1-|S_{11}|}{\sqrt{|S_{11}|}} \tag{4.10}$$

Figure 4.4 shows the proposed antenna configuration and feeding techniques and the PRS unit-cell geometry. This antenna geometry is explored for achieving high gain over the wide frequency range. For polarization-independent applications this type of antenna can be used.

In [7], a frequency-selective surface (FSS) superstrate layer proposed to improve gain obtained an 8.5-dBi gain when the FSS was loaded on the DRA. For the best antenna characteristics, the superstrate size and the air gap height between the antenna and superstrate layer are optimized. This antenna has good potential for millimeter-wave wireless communication and imaging system applications. In [8], a planar metamaterial superstrate loaded slot-coupled HDRA attained an 11.43-dBi gain. Similarly, in [9–11], respectively, a doubled-sided metamaterial loaded compact RDRA [9], a wideband circularly polarized low-

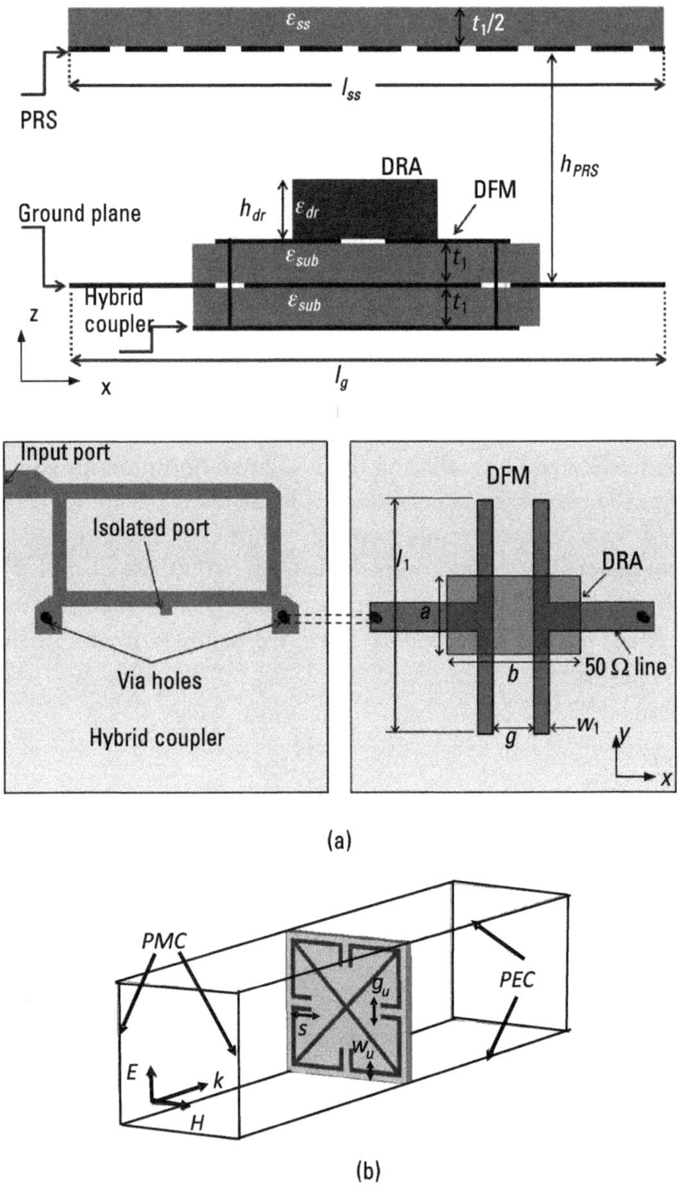

Figure 4.4 PRS-loaded DRA: (a) proposed antenna configuration and feeding techniques and (b) PRS unit-cell geometry [6].

profile DRA [10], and an aperture-coupled biodegradable star-shaped DRA [11] are investigated for enhancing gain.

In [9], a proposed antenna is designed by arrangements of 50-unit cells of metamaterial (MTM). To cover the maximum bandwidth and improve the gain of the RDRA, each unit cell is designed of two parallel eight-shaped conducting strips etched over both faces of a dielectric substrate to offer a negative refractive index of 7.3–8.1 GHz. An extracted lumped circuit model of a unit cell leads to concurrence of EM simulations results. Figure 4.5 shows the MTM unit cell boundary with an equivalent circuit model. The equivalent circuit of the unit cell consists of two symmetrical series LC [capacitance (C) and inductance (L)] resonator circuits and is connected with coupling capacitance (C_g). The capacitors C_s and C_g represent the metallic strip gap and dielectric gap, respectively, whereas the inductors (L_1 and L_2) represent the thin conducting strips that account for the dominant inductive behavior when their effective length is much less than $\lambda/4$. The E field polarization of the incident wave is normal to the gap, and then a surface current is generated in the conducting loop. A resonance similar to that of the LC resonant circuit is achieved when the electrical energy stored in the capacitive gap and magnetic energy stored in the conducting loop are equal. By loading MTM as a superstrate, the proposed RDRA attains a 14-dBi peak gain and a 16.1% (7.18–8.44-GHz) impedance bandwidth.

To improve gain, reflective surfaces, such as a superstrate and bottom substrate, are incorporated into the proposed antenna; a metallic surface or dielectric slab is used for designing a superstrate or bottom substrate to enhance the antenna gain. In addition, various reflector techniques are used to improve the gain of the antenna, such as using a nontransparent conducting sheet as the superstrate resonant cavity in the antenna [12], a CDRA using a flat reflector plane [13], a DRA using a bottom-side reflector [14], and a conical DRA with a superstrate and bottom-side reflector [15]. A nontransparent metal solid sheet as superstrate is deployed in a CDRA to attain a 12-dBi gain over the bandwidth [12]. The CDRA is excited by a probe-feeding technique, and the height of the superstrate position is calculated by a Fabry-Pérot cavity resonating at f_0 by

$$h_0 = \frac{nc}{2f_0} = \frac{n\lambda_0}{2} \tag{4.11}$$

Figure 4.6 depicts a fabricated prototype antenna. The construction of the antenna uses Emerson Cuming's HiK dielectric material ($\varepsilon_r = 10$), and radius

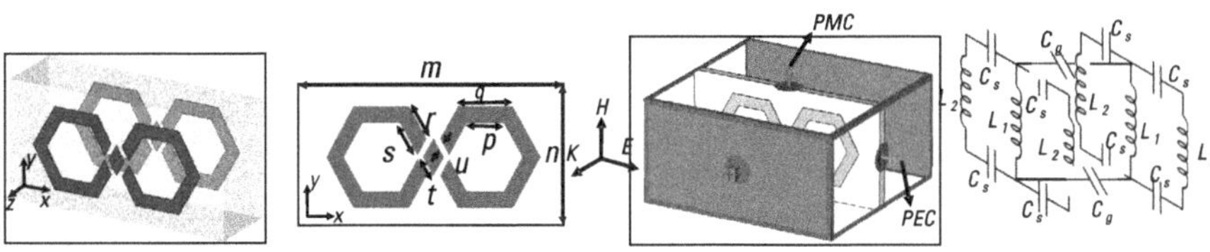

Figure 4.5 MTM unit cell boundary with an equivalent model circuit.

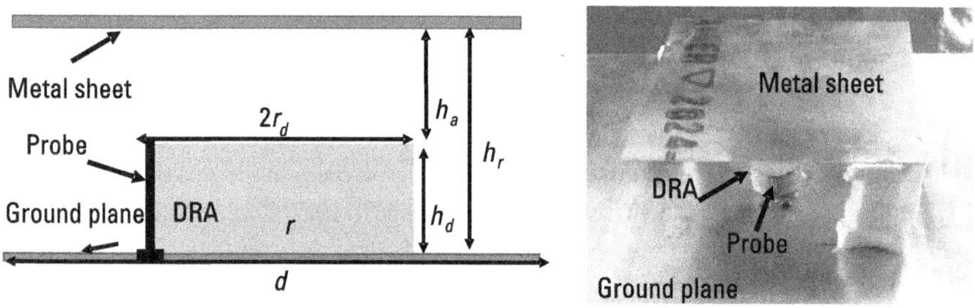

Figure 4.6 Antenna design elements and fabricated prototype [12].

(r_d) = height (h_d) = 10 mm. The superstrate is constructed from an aluminium sheet of thickness (t) = 0.3 mm.

In [15], the gain of a conical-shaped DRA is improved by deploying the superstrate and bottom-side reflector. The reflector below the ground plane mainly focuses on enhancement of the gain (109%), but it also helps to reduce the back radiation. The superstrate maintains a wide frequency band. This antenna attains an 11.25-dBi maximum gain. Figure 4.7 shows a perspective view of the proposed antenna with the superstrate and reflector plane [15].

In [16], a DRA loaded with partially reflective surfaces is proposed for high gain achievement. The DRA is excited by a microstrip line, and a modified stepped ring cross-slot is used to generate CP. Four modified parasitic metallic plates are sequentially fixed around the DRA for improving the AR bandwidth. Later, a partially reflective surface is incorporated in the DRA for enhancing the gain and CP bandwidth. The proposed DRA offers a 54.3% impedance bandwidth and a 54.9% AR bandwidth. A 10.7- dBic and 14.2-dBic average and maximum gain are attained within the frequency range. Figure 4.8 shows the configuration of the proposed antenna, which is surrounded by four metallic plates to generate strong coupling and, therefore, the higher-order modes that deliberately increase the gain.

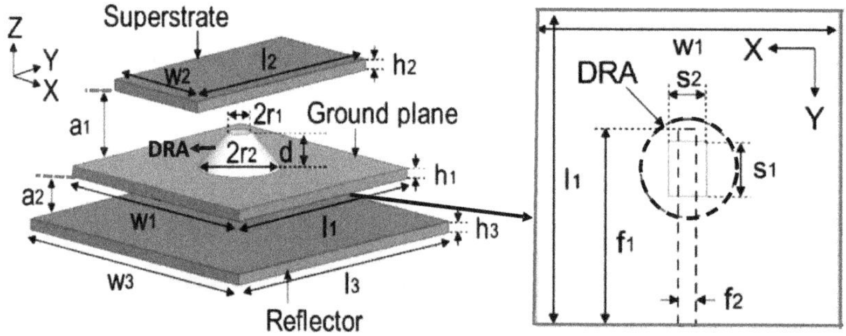

Figure 4.7 Proposed antenna perspective view with superstrate and reflector plane [15].

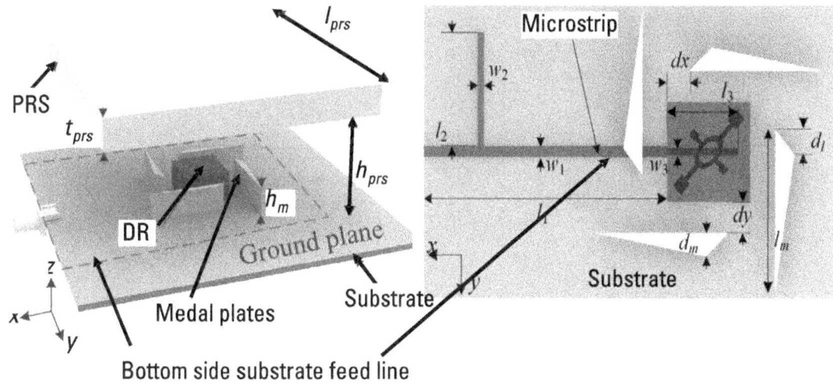

Figure 4.8 Configuration of the antenna geometry proposed in [16].

4.2 Gain Enhanced by Metal Integration of the DRA

This section details the use of new techniques that achieve high gain with wide bandwidth, such as metal integration with substrate, DRAs surrounded by metallic horns, metallic cap loading, and edge grounding [17-24]. Figure 4.9 shows numerous metal intergradation techniques that improve gain.

In [17], a hybrid DRA with a circular mushroom-like structure is investigated for enhancing gain. A cylindrical metallic electromagnetic bandgap substrate is used to design a CDRA with improved performance in antenna attributes. The CDRA (depicted in Figure 4.10) is excited by a metallic probe and integrated within a cylindrical metallic periodic electromagnetic bandgap (EBG) substrate to increase the antenna gain to 9.84 dB. Two different types of periodic structures are used to make the EBG structure. One element consists of metal rings and the second element of vertical grounding metal vias, which are arranged at certain distances to form radial and circular periodic structures. The concentric rings of the strips are printed on the grounded plane of substrate.

For maximum gain, two radial periods P_{r1} and P_{r2} are selected at different positions, close to a quarter-wavelength distance. Due to the circular symmetry of the proposed EBG structure, the capacitive coupling is the same in all radial directions. The grounded substrate consists of a dielectric material ($\varepsilon_r = 2.2$),

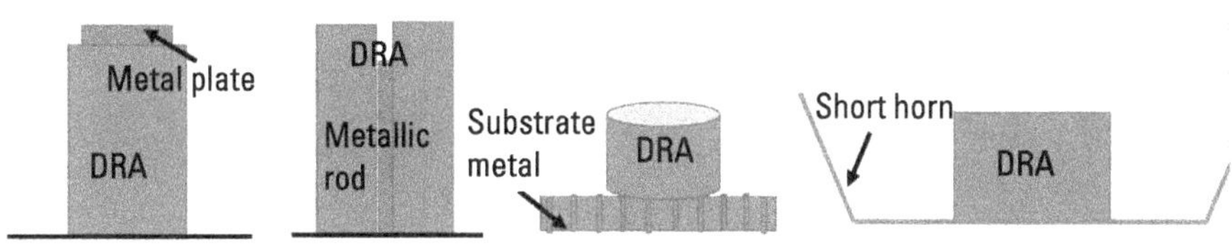

Figure 4.9 Metal intergradation techniques: (a) metal plate loading, (b) metallic rod insertion, (c) substrate metal integration, and (d) peripheral outside the metal boundary (short horn).

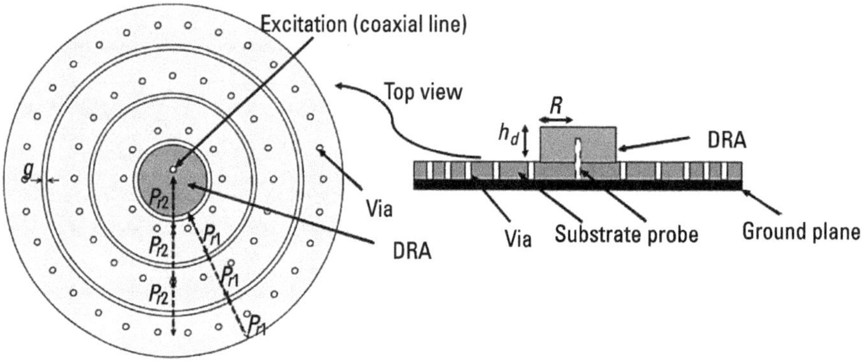

Figure 4.10 CDRA surrounded by a metallic circular mushroom-like substrate [17].

and the CDRA is constructed from a high-permittivity dielectric material (ε_r = 35.5).

The hybrid DRA technique used in [18] aims to achieve a high-gain millimeter-wave application. The DRA (ε_r=10) is excited by a circular microstrip patch-feeding network that is printed on a low-dielectric substrate (ε_r= 2.2). An optimized feed length and position excites a higher-order mode ($HEM_{15\delta}$), which helps to improve the radiation of the antenna. The proposed antenna configuration achieved an 11.9-dB gain by coupling a microstrip patch with a ring-shaped DRA to increase the electrical area of the antenna. Figure 4.11 shows a fabricated prototype of the antenna, which obtained a measured gain of 11 dB from 55 to 63 GHz, which is enough to cover the whole ISM band.

In [19], a substrate integrated waveguide (SIW) with a DRA antenna system is designed to improve antenna performance to meet millimeter-waveband applications' demand for high gain combined with high radiation efficiency. The proposed SIW structure consists of two integrated rows of metallized vias and separated by (SIW width). The position of the slot on an SIW controls the amount of coupled power to the DRA. For the fundamental mode, the

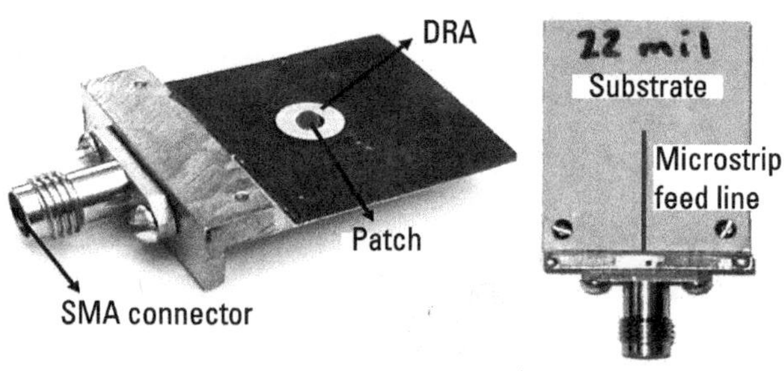

Figure 4.11 Fabricated prototype of an antenna with a microstrip feed [18].

maximum excitation to the RDRA requires a slot, which is situated at the center of the SIW. Figure 4.12 depicts the SIW-based DRA configuration, which achieved a 5.51-dBi measured gain in the broad side direction.

Similarly, in [20], metallic vias loaded differentially coplanar-fed filtering DRA is designed for high gain performance. In [21], a rectangular filtering DRA is investigated for a wide band and high gain. It is excited by a microstrip-coupled slot from a bottom-side substrate, and the open stub of the microstrip feed line is designed to provide two radiation nulls at the frequency band for a filtering function. Two metallic parasitic strips are added in parallel to the microstrip feed line to provide good suppression in the frequency band. A compact filtering DRA with a metallic parasitic configuration offers a quasi-elliptic bandpass response without involving specific filtering circuits. A further 4-dBi gain is achieved when the modified DRA is fed by a pair of separated slots. Figure 4.13 depicts the configuration of this filtering DRA, which offers a 10-dB impedance bandwidth of 20.3% and an average gain of 9.05 dBi within the passband.

In [22], a metallic boundary configuration integrated with DRA is used to amplify radiation in a particular direction to improve the gain for wideband

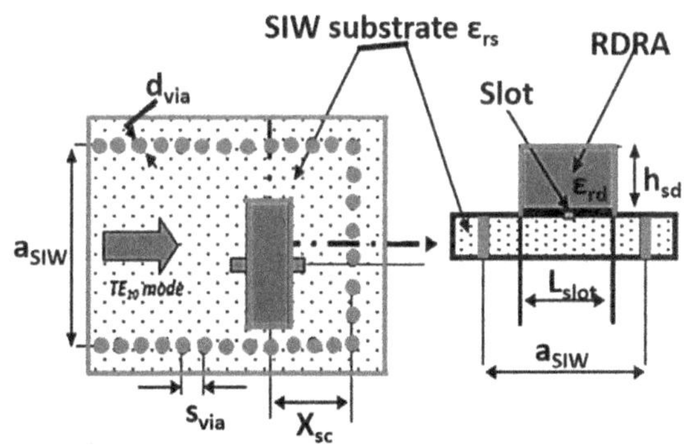

Figure 4.12 SIW-based DRA configuration [19].

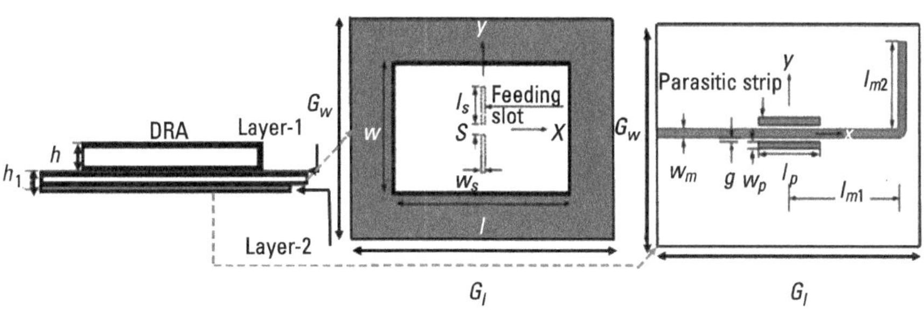

Figure 4.13 Configuration of filtering DRA with enhanced gain [21].

wireless applications. To design the antenna, a low-dielectric material is used to construct the hollow CDRA with a top-mount spherical cap lens, and a metal reflector is used toward the outer side of the antenna geometry. It is excited by a metallic probe-feeding network. The convenient shaping of the lens and metal reflector outside the ground boundary yields a high gain (14 dBi) and minimized back radiation. Figure 4.14 shows this optimized antenna geometry configuration. The hollow CDRA consists of an inner radius $R_{\mathrm{dri}} = 5$ mm and an outer radius $R_{\mathrm{dre}} = 10.20$ mm, and its height $h_{\mathrm{dr}} = 19.3$ mm, with a dielectric constant $\varepsilon_{r1} = 3$. The spherical cap lens has a radius $R_{\mathrm{l}} = 32$ mm; the opening angle $\varphi_{\mathrm{l}} = 144°$ is designed from a dielectric constant $\varepsilon_{r2} = 4$ material. The DR is fixed at the center of a circular ground plane metal reflector of radius $R_{\mathrm{sc}} =$ 29.8 mm. This particular shape of the antenna reflector geometry limits the backscatter radiation because of the edge currents produced along the metal reflector plate. Furthermore, a metal reflector peripheral plate is designed not only to increase the gain, but also to drastically reduce the backscattering by exploiting the double-edge diffraction phenomenon.

Similarly, in [23], a bent ground plane with a DRA is designed to achieve a 5.6-dBi gain. The radiation beam width is increased to more than 120° by bending the metallic ground plane. In addition, four metal plates are attached to the edge of the ground plane to enhance gain in the broadside direction. The wide return-loss band accomplished a 3.4–3.7-GHz bandwidth by merging

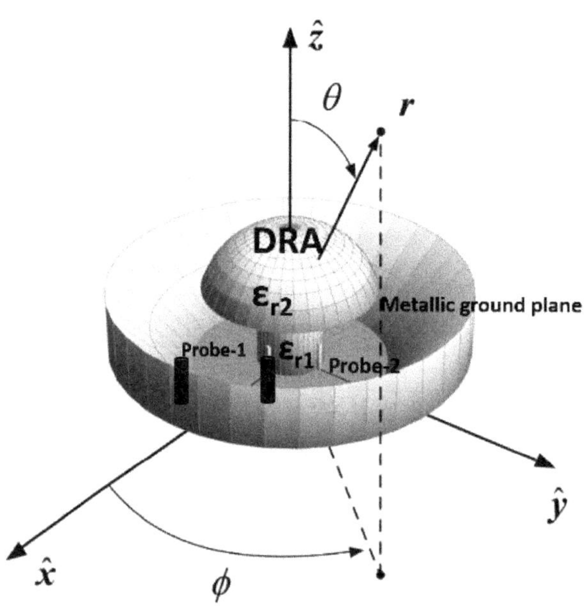

Figure 4.14 Optimized antenna geometry configuration [22].

bands. Figure 4.15 shows the geometry of the DRA with four plates attached with a ground plane.

In [24–27], various dielectric resonators integrated with mounted horn antennas are investigated for enhancing the gain. In [24], a low-profile with a compact cross-dielectric resonator on a microstrip patch (DRoP) and a quasi-planar surface-mounted short horn (SMSH) achieves a gain up to 9 dBi [24]. The antenna consists of a cross-DR aperture coupled to a microstrip patch, and its ground plane surface is mounted on short horn. This antenna geometry is capable of attaining a wide-impedance bandwidth and a high gain due to its short-horn integration. For enhancing the gain of the DR with a microstrip patch, the height of the SMSH is erected almost at the height of the DRoP. By choosing the appropriate optimized dimensions and height of the SMSH, an increased radiation aperture and gain over the impedance bandwidth can be controlled significantly. Figure 4.16 presents this proposed DRA configuration with a fabricated prototype.

Similarly, differentially feeding an RDRA with a short antenna achieves an optimum boresight gain of 12.2 dBi [25]. 3% to 30% of bandwidth is improved by loading stub in the feeding network. The surface-mounted short horn is made of lightweight aluminium material. In [26], a CDRA integrated with a conical horn antenna is proposed for high gain. The aluminum-based conical-shaped horn is surrounded with an inverted CDRA to achieve a maximum 14.1-dBi peak gain at a 10.7-GHz resonant frequency. The antenna maximum gain is achieved by optimizing the aperture angle, $\theta = \tan^{-1}\left(\frac{r_h}{h_h}\right)$. The best results are obtained when the phase error of the horn is at a minimum, as

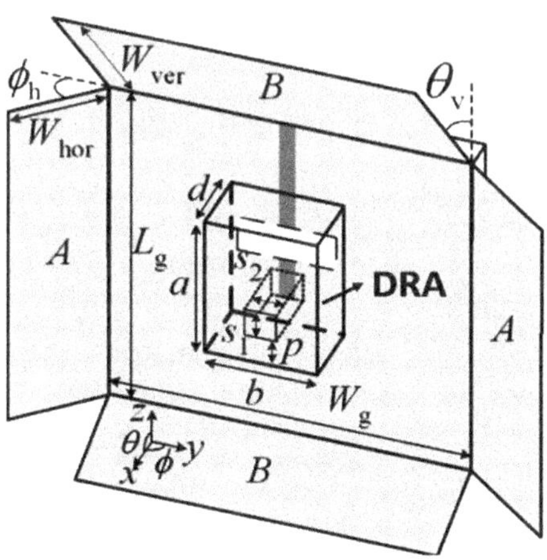

Figure 4.15 Geometry of DRA with four plates attached with ground plane [23].

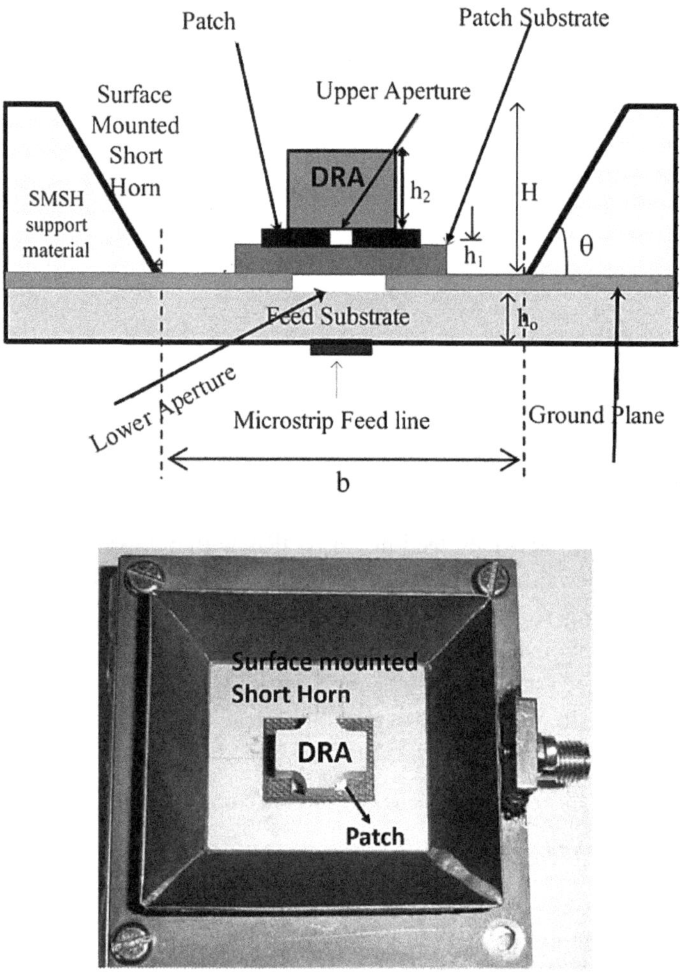

Figure 4.16 Proposed DRA configuration with a fabricated prototype of a short horn [24].

the phase error increases with the flaring of the aperture angle (θ). Figure 4.17 shows the fabricated conical-shaped horn with a DRA configuration and a radiation pattern at 9.2 GHz.

Another new technique for achieving high gain with a CDRA involves using an intermediate layer and a metallic cylinder [28]. The intermediate layer structure can abolish the surface wave propagation and hence improve the gain. In addition, the cavity formed by the metallic cylinder that is surrounded by the CDR enhances the gain. The proposed antenna achieves an operating bandwidth of 23% from 5.4 to 6.8 GHz with an 11-dBi high gain. Figure 4.18

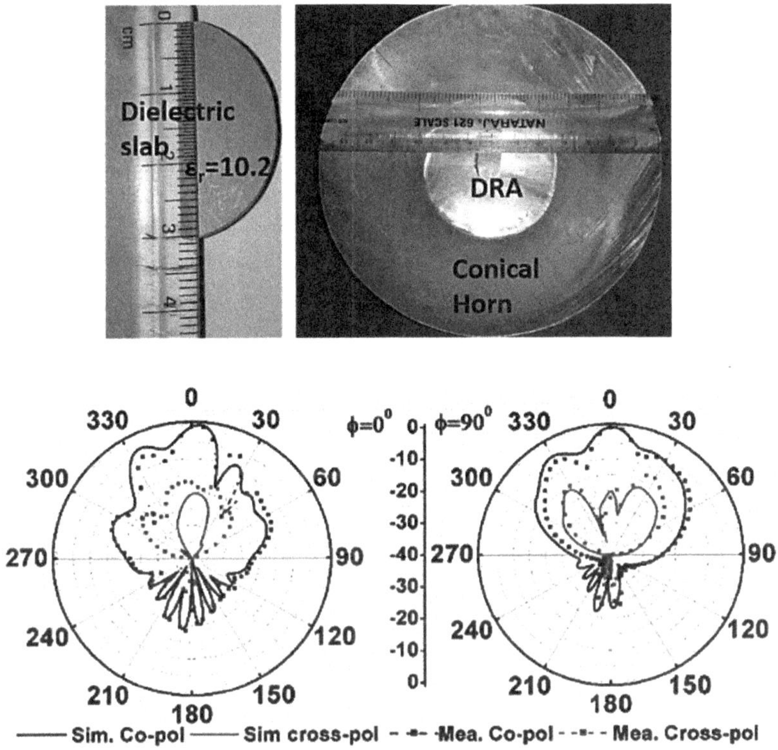

Figure 4.17 Fabricated conical-shaped horn with a DRA configuration and a radiation pattern at 9.2 GHz [26].

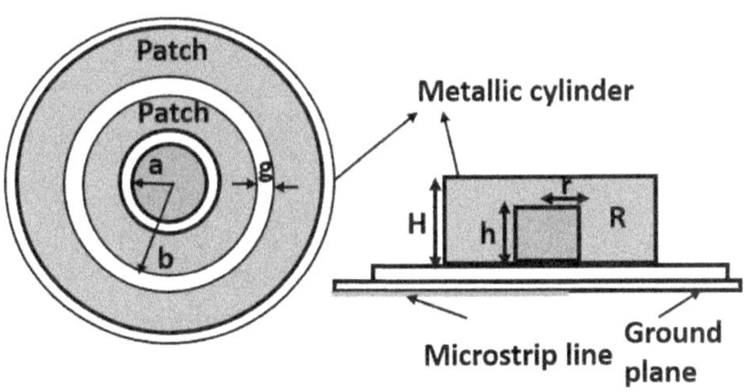

Figure 4.18 Structure of the antenna with an intermediate layer and a metallic cylinder proposed in [27].

shows the structure of the proposed antenna. Similarly, [29] discusses a CDRA with a double-annular patch and a metallic cylinder that is used for obtaining a large bandwidth (5.3–6.8-GHz) and a high gain (11-dBi).

In [30], a stacked CDRA loaded with a metallic patch is investigated for high gain. Two techniques are considered: the stacking of two dielectric materials, which enhances the bandwidth of the antenna (to 5.7–13.2-GHz), and using a metallic patch as a reflector, which improves the gain up to 6.8 dBi and keeps the antenna compact. In [31], the miniaturization of the DRA by stacking and metallic cap–loading are considered to improve gain. Here, the compactness of the antenna is achieved by choosing an appropriate dimension of the metallic cap. The antenna is operated at a lower frequency range and enhances realized gain. Furthermore, the wavelength at the *z*-axis improves as the radius of the metallic cap increases and diminishes in the lower frequency range. In addition, the tangential field at the side wall surface improves, resulting in an increase in realized gain at the resonant frequency. Figure 4.19 illustrates the geometry of the proposed antenna, its simulated impedance bandwidth, and its gain plot. At a 5.9-GHz resonant frequency, a 8.54-dBi gain is achieved by metallic cap–loading.

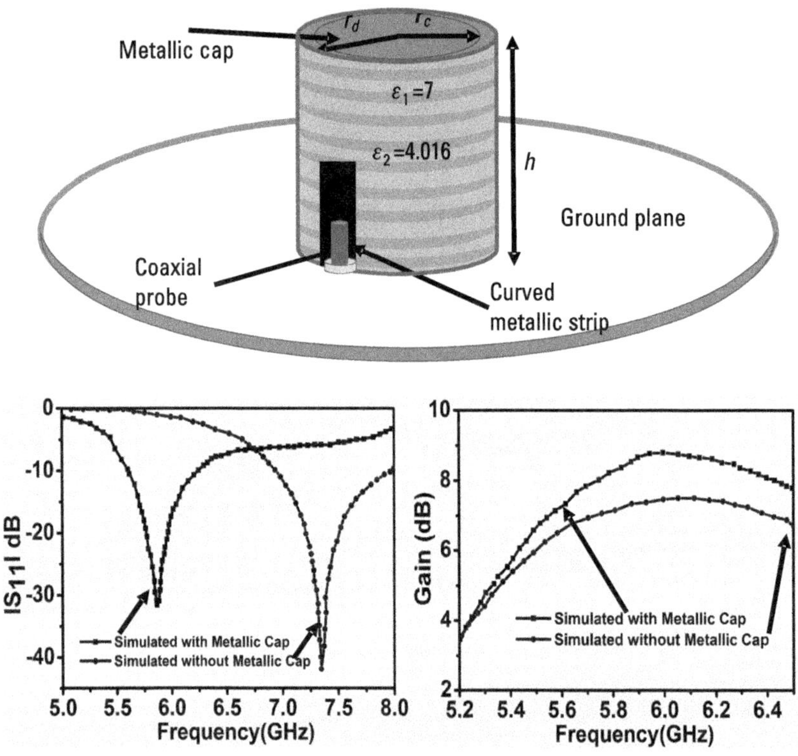

Figure 4.19 Stacked CDRA: (a) geometry of the proposed antenna and (b) simulated impedance bandwidth and gain plot [31].

4.3 Gain Enhanced by Higher-Order Modes

This section discusses the use of exciting various higher modes to enhance the gain of the antenna. In some of these cases, a more than 3-dBi gain /directivity in a single-element DRA can be achieved by exciting a higher-order mode. The higher-order modes of DRAs have already been used in multimode operations to improve the impedance bandwidth performance. Numerous modes with different antenna characteristics and radiation patterns have been excited in a DRA. Apart from the fundamental mode, various higher-order modes can be generated in a DRA through different feeding techniques and ARs of the modes, such as $TE^x_{\delta n1}$ and $TE^x_{\delta 1n}$, $HE_{11\delta}$ [32]. Nevertheless, not all the higher-mode radiation patterns are similar to the fundamental modes of radiation patterns (e.g., the broadside pattern or the null at the broadside).

Several higher modes have been proposed to enhance the gain of DRAs [33–42]. Higher modes are also deployed to improve the impedance bandwidth in [43–44]. In [33], higher-order modes of an RDRA are rigorously investigated for achieving higher gain or directivity. A dielectric waveguide model is used to calculate the resonant frequency of the higher-order modes. The RDRA excites the $TE^x_{\delta 15}$ mode and enhances the directivity up to 11 dB as compared to the lowest-order mode with 5 dB. Figure 4.20 presents various higher-order modes and their equivalent magnetic dipole models. The spacing between the magnetic dipoles of the higher-order modes will affect the radiation pattern of the DRA. In Figure 4.21, the radiation pattern depicts the far-field patterns of the DRA at the fundamental $TE^x_{\delta 11}$ and higher-order modes $TE^x_{\delta 13}$ and $TE^x_{\delta 15}$ for spacing between magnetic dipoles ($s = 0.4\lambda$).

The DRA proposed in [33] is designed with dielectric material $\varepsilon_r = 10$ and dimensions $w = d = 5$ mm and $h = 30$ mm. Following the transcendental equation (4.12) for the $TE^x_{\delta 15}$ mode ($m = 1$, $n = 5$), it is found that antenna resonates at 11.2 GHz. The dielectric resonator is mounted on a 100-mm × 100-mm ground plane and excited by aperture microstrip feeding techniques.

$$k_x \tan\left({}^{k_x d}\!\big/_{2} \right) = \sqrt{\left(\left(\varepsilon_r - 1 \right) k_{mn}^2 - k_x^2 \right)} \tag{4.12}$$

Similarly, an RDRA with high gain is investigated in [34]. To obtain the desired results, various higher modes and the aspect ratio of the DRA design configuration are analyzed. Figure 4.22(a) shows the three different aspect ratio dimensions of DRA and the correspondingly excited various higher-order modes. The higher-order modes significantly narrow the beamwidth of antenna, which indicates an increase in directivity. In addition, adjusting the aspect ratio controls the radiation in the desired direction. With the proposed

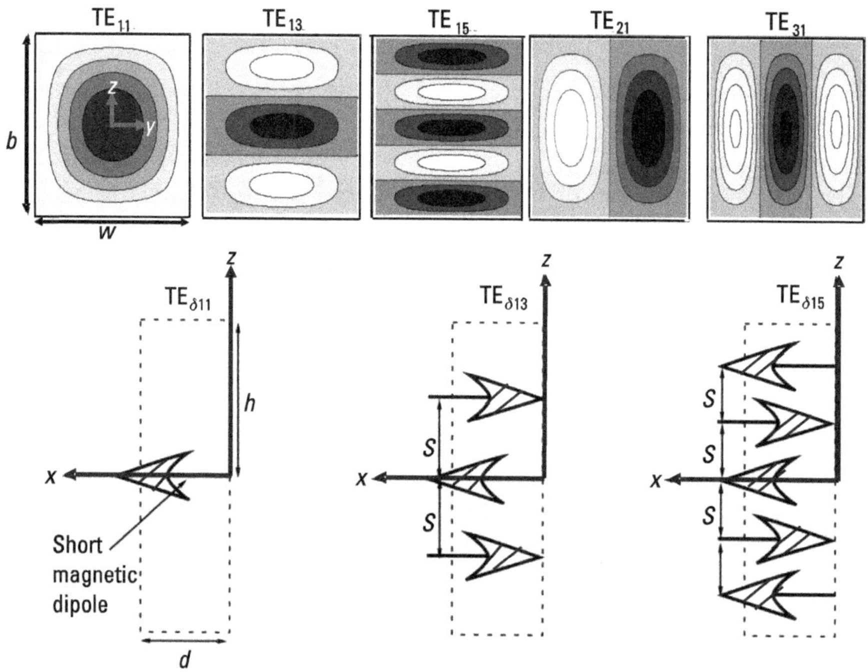

Figure 4.20 Various higher-order modes and equivalent magnetic dipole models in an RDRA [33].

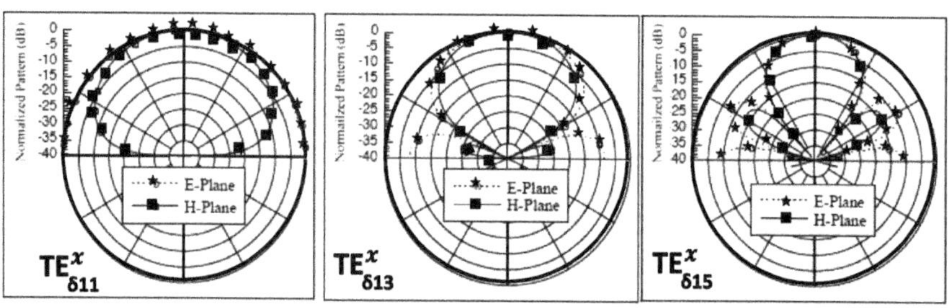

Figure 4.21 Radiation patterns of the DRA at the fundamental $TE^x_{\delta 11}$ and higher-order modes $TE^x_{\delta 13}$ and $TE^x_{\delta 15}$ [33].

RDRA, a 10.2-dBi gain is achieved at the $TE^x_{\delta 15}$ operating mode. Figure 4.22(b) illustrates the gain plot for all different aspect ratios of the antenna.

A higher-order mode RDRA for millimeter-wave applications is investigated in [35]. The higher-order mode RDRA is excited by a slot-coupled microstrip feeding network. It is observed that when radiating slot is centrally located below the DRA, a TE_{pqr} mode can be generated only when all of the indices *p*, *q*, and *r* are odd numbers. The optimized aspect ratio of the DRA offers a single TE^y_{11r} higher-order mode. For the TE^y_{115} and TE^y_{119} higher-order

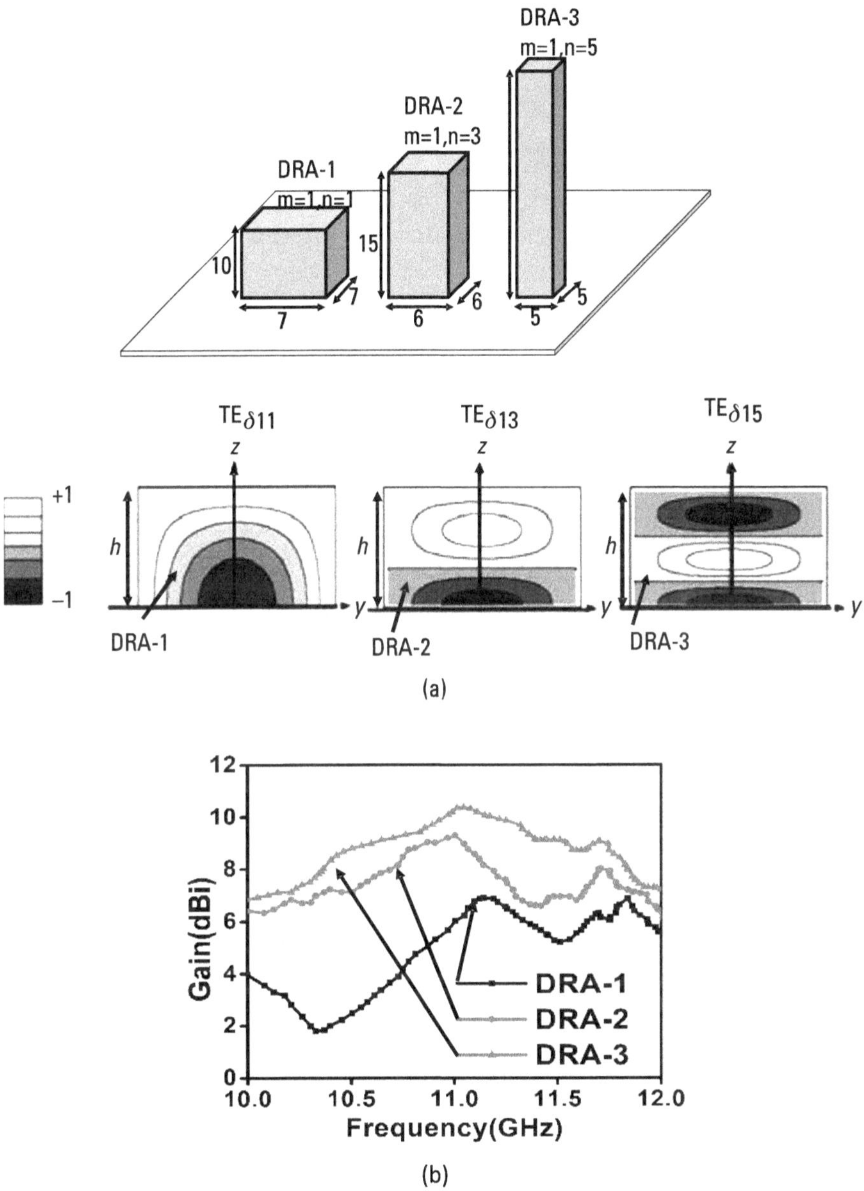

Figure 4.22 (a) Three different aspect ratio dimensions of DRA and various excited higher-order modes and (b) gain plot [34].

modes, the geometries of two different DRAs were fabricated by an Eccostock HiK dielectric material ($\varepsilon_r = 10$). The microstrip feed line was etched on a Duroid substrate ($\varepsilon_{rs} = 2.2$) and a thickness of $h = 0.25$ mm. For the TE^y_{115} and TE^y_{119} modes, respectively, a 5.8-dBi and 6.3-dBi gain are obtained at a 24-GHz resonant frequency. Also, the operating frequency shift is dependent on the aspect ratios of the DRA. Figure 4.23 shows the proposed DRA structure and frequency percentage shifting of the aspect ratio (*b*/*d*) for TE^y_{111}, TE^y_{115}, and TE^y_{119} modes [35].

Similarly, in [36], a higher-order DRA realizes high gain for the 340-GHz frequency. The antenna consists of a low-loss dielectric resonator that consists of a high-resistivity silicon material and an on-chip feeding-patch complementary metal-oxide-semiconductor (CMOS) technology for exciting the desired modes. $TE_{\delta 17}$, a higher-order mode, is excited, significantly improving the antenna gain to 6.7dBi. Moreover, the antenna ground plane protects the EM fields from leaking into the lossy CMOS silicon substrate, improving the antenna's performance. Figure 4.24 shows the proposed high-gain terahertz antenna and its higher-order mode.

In [37], a two-layer circularly polarized $TE_{11,11}$ RDRA is considered for high gain. The design configuration incorporates the DRA and a dielectric coat layer with their dielectric constants (ε_r= 10 and 3.5). The outer layer offers a wider impedance with CP bandwidths of ~21% and 9.5%, respectively, and a high gain of ~11 dBic. Figure 4.25 depicts the configuration of the antenna and excited H field at 11.3 GHz for $TE_{11,11}$. The DRA inner magnetic field distribution is represented as an array of short magnetic dipoles, in which a

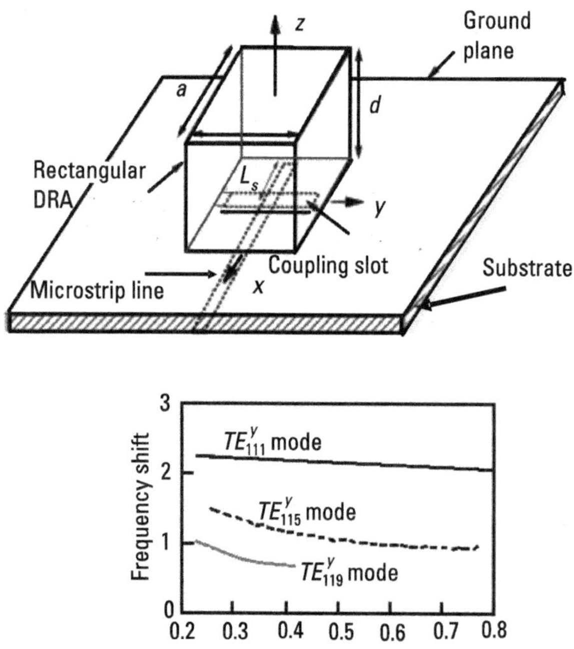

Figure 4.23 Proposed DRA structure and frequency percentage shifting of the aspect ratio (b/d) for TE^y_{111}, TE^y_{115}, and TE^y_{119} modes [35].

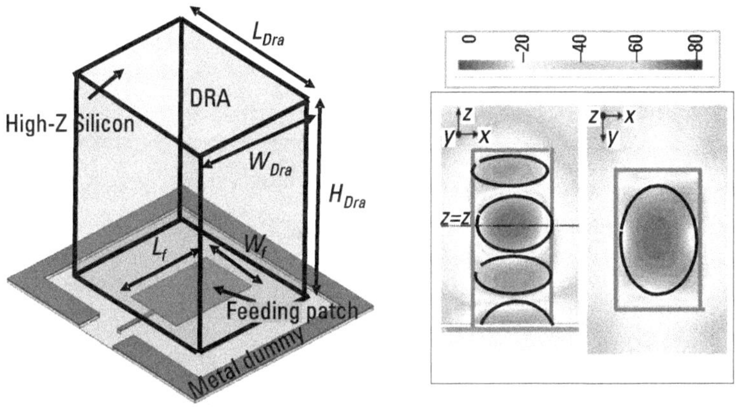

Figure 4.24 High-gain terahertz antenna and its higher-order mode [36].

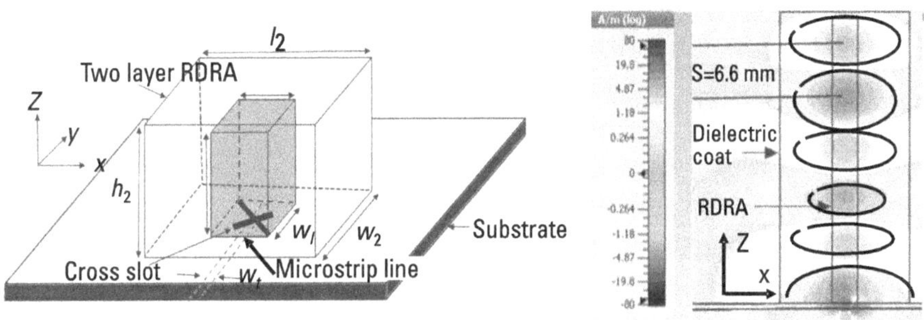

Figure 4.25 Configuration of antenna and excited H field at 11.3 GHz for TE11,11 [37].

maximum gain can be accomplished when the distance between the adjacent magnetic dipoles is ~0.4λ.

A higher-order-mode excitation for high-gain broadside radiation from a CDRA is explored in [38]. In a CDRA, resonant modes such as $HEM_{12\delta}$, $HEM_{11\delta}$, and $TM_{01\delta}$ are examined for various radiating applications. The most challenging mode excitation is explored and overcome by using an inventive technique; specifically, an $HEM_{12\delta}$ mode excitation with high-gain broadside radiation is experimentally demonstrated. If a metal sheet or electric boundary is incorporated across the ground plane, the E fields across the horizontal ground plane vanish immediately and destroy the boundary condition to excite the mode. Hence, a conventional ground plane does not excite the $HEM_{12\delta}$-like mode. Figure 4.26 shows a schematic diagram of the proposed antenna and the $HEM_{12\delta}$, $HEM_{11\delta}$, and $TM_{01\delta}$ E field distributions. The proposed antenna configuration [38] achieves a peak gain in broadside radiation of more than 8 dBi.

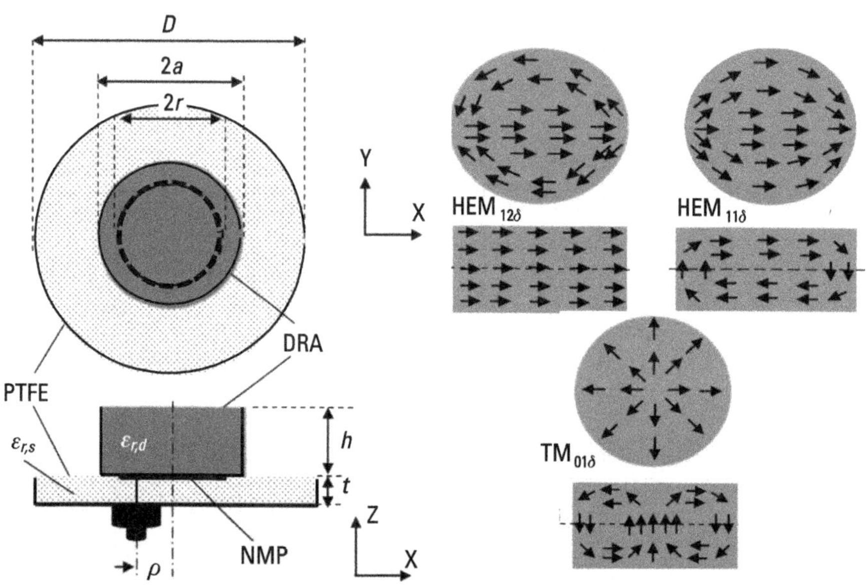

Figure 4.26 Schematic diagram of the proposed antenna and $HEM_{12\delta}$, $HEM_{11\delta}$, and $TM_{01\delta}$ E field distributions [38].

In [39], the authors use a composite aperture to excite two different modes, $HEM_{12\delta}$, $HEM_{11\delta}$, in a CDRA. Specifically, they introduce a new feeding technique that generates an electric current and induces the $HEM_{12\delta}$ mode in a CDRA. The proposed CDRA is designed with an optimized aspect ratio (height/radius) and dielectric material ($\varepsilon_r = 10$). The feeding structure is incorporated on a low-dielectric substrate $\varepsilon_{rs} = 2.33$. At a 3.9-GHz and 7.4-GHz resonant frequency, respectively, the $HEM_{11\delta}$ and $HEM_{12\delta}$ modes are excited. Figure 4.27 shows the apertures for their respective modes. A maximum gain of 6.9 dBi is achieved at a 3.9-GHz resonant frequency.

In [40] a combination of two higher-order hybrid EM modes (HEM_{113} and HEM_{123}) in a CDRA is examined to enhance the antenna gain [40]. An aperture slot in the ground plane of a microstrip line excites the higher-order mode and increases the gain to 11.6 dBi. A 2.6% impedance bandwidth for the ISM band is covered by the antenna. For fabrication, a layered Arlon AR600

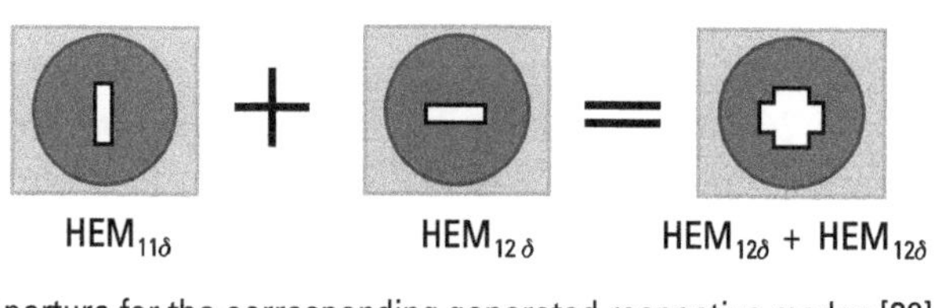

Figure 4.27 Aperture for the corresponding generated respective modes [39].

substrate of relative permittivity 6.15 is used. An aperture-coupling-feeding technique minimizes the excitation of unwanted lower-order modes in the CDRA structure. Figure 4.28 depicts the simulated E-field and H-field distributions of the desired modes in several planes of the resonator at a 5.8-GHz (HE_{133}) resonant frequency.

In [41], the authors describe psi (Ψ)-shaped microstrip feeding techniques through the $HEM_{11\delta}$, $TM_{01\delta}$, and $HEM_{12\delta}$ radiating modes. The proposed CDRA covers 2.5–3.02-, 3.76–3.86-, and 4.38–4.72-GHz, three different frequency bands respectively for WiMAX (2.5 GHz) and vehicular applications. A maximum gain of 7.1 dBi corresponding to the $HEM_{12\delta}$ mode is achieved. Similarly, a pattern-reconfigurable higher-order mode spherical DRA achieved a 9.12-dBi gain.

In [42], a magnetic-dipole quasi-Yagi antenna using a higher-order-mode ($TE_{\delta 31}$) dielectric resonator with near-zero-index metamaterial is investigated for high gain. The higher-order $TE_{\delta 31}$ mode can be used as a magnetic-dipole driver to design a quasi-Yagi antenna for high gain. To improve the antenna gain further, a near-zero-index (NZI) metamaterial (NZIM) is used, instead of the traditional directors, and put in the front of DR driver. With the higher-order mode operation of the DRA and NZIM, a 10.3-dBi gain is realized in the proposed antenna. Figure 4.29 shows the antenna's fabricated prototype and its simulated E fields at 10.55 GHz with and without NZIM.

4.4 Enhancing Gain with a Dielectric Resonator Array System

A DRA arranged in a specific order (or array) can achieve high gain and a radiation pattern that can cancel the interference from undesired directions for long-distance wireless communication. The spacing between the elements, the arrangement of the antenna, and the feeding techniques are important attributes of such an antenna array system, which can improve such factors as directivity, radiation in a specific direction, and SNR.

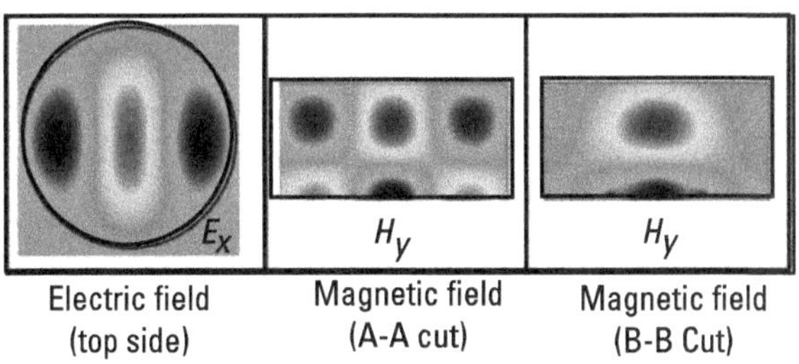

Figure 4.28 Simulated E-field and H-field distributions of the desired modes in several planes of the resonator at a 5.8-GHz (HE133) resonant frequency [40].

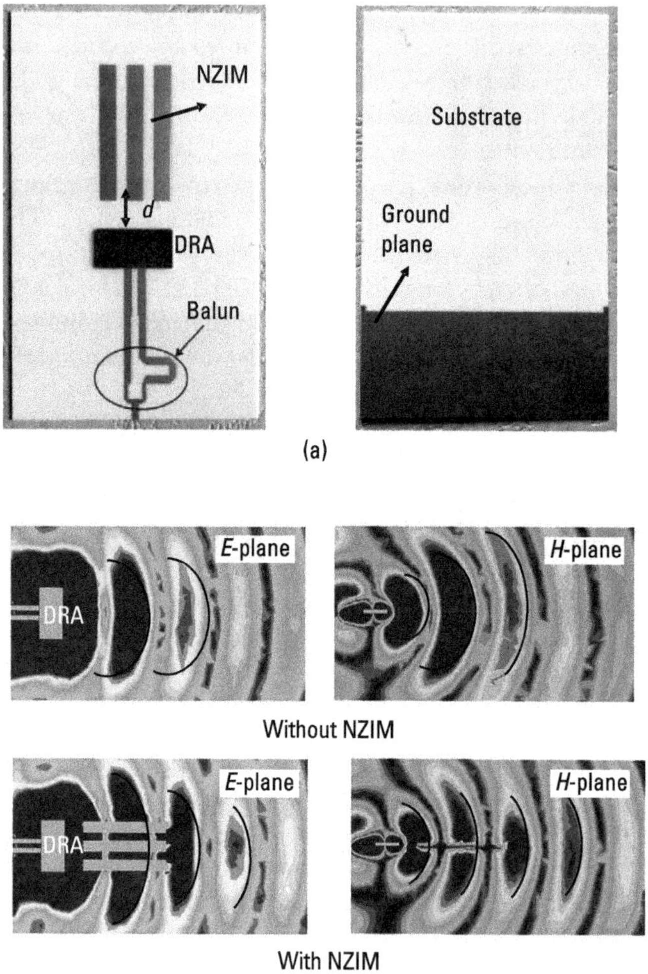

Figure 4.29 Metamaterial and DRA: (a) fabricated prototype of the antenna and (b) simulated E fields and H fields at 10.55 GHz with and without NZIM [42].

- *Theory of antenna array arrangements [1]:* The total radiation pattern (RP) of an array of N elements that is situated at a location (θ, φ) in the far-field region, can be approximated as a product of each radiation intensity (RI) of antenna element and array factor (AF) such that

$$\mathrm{RP} = 20\log\left(RI \times AF\right) \tag{4.13}$$

where $\mathrm{AF} = \sum_{n=1}^{N} A_n e^{j\Psi_n}$ and $= \Psi_n = k_0 x_n \sin\theta \cos\varphi + k_0 y_n \sin\theta \sin\varphi$.

k_0 is the free-space wave number, (x_n, y_n) is the position of the nth antenna element, and A_n is the feeding voltage excitation. Mostly, the AF depends on the arrangement of the antenna array lattice and the excitation feeding techniques.

- *Linear array arrangement:* An N number of array elements with a uniform amplitude. $(A_1 = A_2 .... = A_n)$ and spacing (d) and linear progressive phase shifts $(\beta_1 = \beta_1 ... = \beta_n)$ simplify to an array factor as given in

$$AF = \sum_{n=1}^{N} A_n e^{jnk_0 d\, sin\theta} \tag{4.14}$$

$$AF_n = \frac{1}{N}\left[\frac{\sin\left(\frac{N\psi}{2}\right)}{\text{N}\sin\left(\frac{\psi}{2}\right)}\right] \tag{4.15}$$

where $\psi = k_0 d\theta \sin + \beta$.

- *Planar array arrangement:* Two-dimensional (2-D) planar element arrays offer more symmetrical radiation patterns with lower sidelobes and higher directivity or gain (narrow main beam).

  $M \times N$ elements are arranged uniformly along a rectangular grid in the xy-plane with element spacing d_x and d_y in the x- and y-directions, respectively. For a uniform amplitude, with progressive phase shifts in each direction, the total AF is calculated as

$$AF(\theta,\phi) = I_0 \sum_{m=0}^{M-1} e^{jk(md_x \sin\grave{e} \cos\phi + \beta_x)} \sum_{n=0}^{N-1} e^{jk(nd_y \sin\grave{e} \sin\phi + \beta_y)} \tag{4.16}$$

the normalized AF

$$AF_n(\theta,\phi) = \left[\frac{1}{M}\frac{\sin\left(M\frac{\psi_x}{2}\right)}{sin\left(\frac{\psi_x}{2}\right)}\right] \times \left[\frac{1}{N}\frac{\sin\left(N\frac{y}{2}\right)}{sin\left(\frac{\psi_y}{2}\right)}\right] \tag{4.17}$$

β_x and β_y is progressive phase shift in each direction.

The number of antenna elements, the element position arrangements, and their excitation feed technique coefficients can be achieved for desired high radiation, directivity, or gain for various wireless applications. The general expression of the directivity of an array is given as

$$D_0 = 4\pi \frac{\left|AF_n(\theta_0,\phi_0)\right|^2}{\int_0^{2\pi}\int_0^{2\pi}\left|AF_n(\theta_0,\phi_0)\right|^2 \sin\theta d\theta\, d\phi} \tag{4.18}$$

where θ_0 and φ_0 are the main beam direction radiation.

- *Circular planar array:* For circular antenna array arrangements, the AF can be represented as

$$AF(\theta,\phi) = \sum_{n=1}^{N} I_n e^{j[ka\sin\theta\cos(\phi-\phi_n)+\alpha_n]} \tag{4.19}$$

where I_n is the amplitude of excitation, $\varphi_n = (2\pi/N)n$ is the angular position of the nth element, and α_n is the phase of the excitation of the nth element. Figure 4.30 shows the various array arrangements.

Various DRA array configurations have been reported [45–55] for improving gain or directivity. In [45], an aperture-coupled RDRA array with parasitic elements is investigated for high gain. Here, the antenna gain is enhanced by controlling the adjacent element spacing. The DRA is excited by a slot aperture–feeding technique. An array of parasitic adjacent DRAs is excited by strong mutual coupling through the main DRA. Two different antenna-array arrangement configurations are investigated for improving antenna characteristics. Initially, parasitic DRs are placed next to the main DRA, which is constructed with different dielectric material or different sizes. The impedance bandwidth

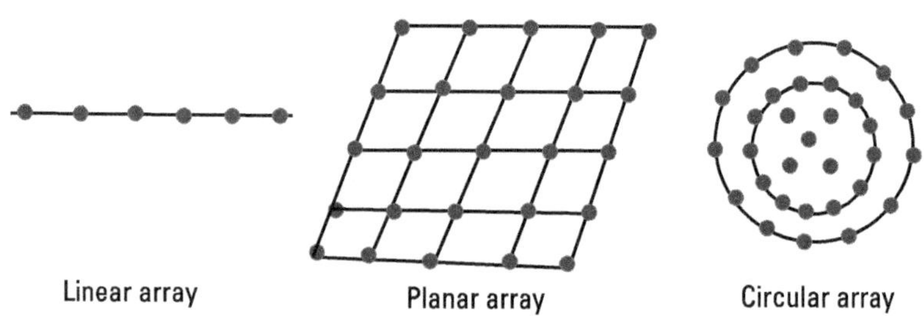

Figure 4.30 The array arrangements of various antenna elements.

increases but degrades the radiation pattern characteristics. In a second antenna arrangement, parasitic elements are placed in a symmetric arrangement (cross-shaped), on four sides of the main DRA, and offering a tilted main beam for the array. Thus, the radiation pattern in the broadside direction is improved by using the parasitic elements on both sides of the main feeding DRA. Figure 4.31 shows both antenna geometry configurations. A significant improvement in gain is obtained when using adjacent element spacing; the seven-element linear parasitic array offers an 11.14-dB gain; and the cross-shaped, five-element parasitic array offers a 10.1-dB gain.

In [46], a steerable dielectric resonator phased-array antenna is reported for enhancing gain. For steering the radiation beam of antenna, an inkjet-printed tunable phase shifter with metal-insulator-metal varactors is used. The DRA is fabricated of a bulk-glass ceramic with a stacked antenna architecture for wide bandwidth and high gain. The proposed antenna consists of a feeding network connected to the four branches of the array. Every branch receives the split-input signal with a phase shifter as the controlling element for each antenna element. Figure 4.32 depicts the layout of the four-element DRA with a feeding network; this proposed array configuration of DRA elements delivers a gain of almost 13 dBi for various wireless applications.

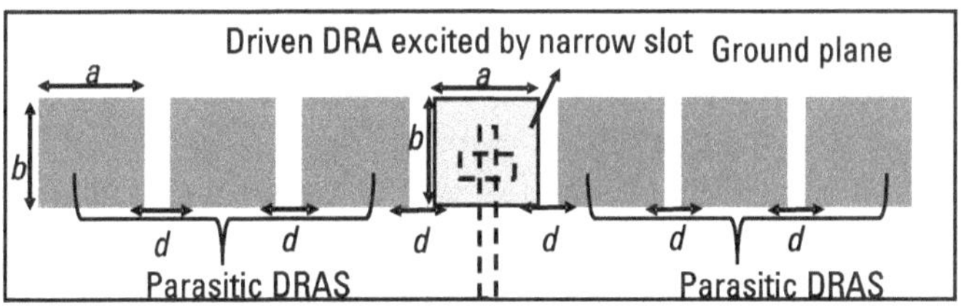

(a)

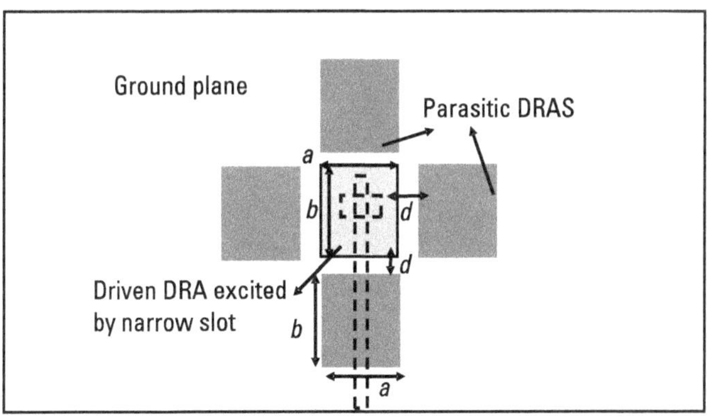

(b)

Figure 4.31 Both antenna geometry configurations: (a) the seven-element linear array and (b) the five-element cross-shaped array [45].

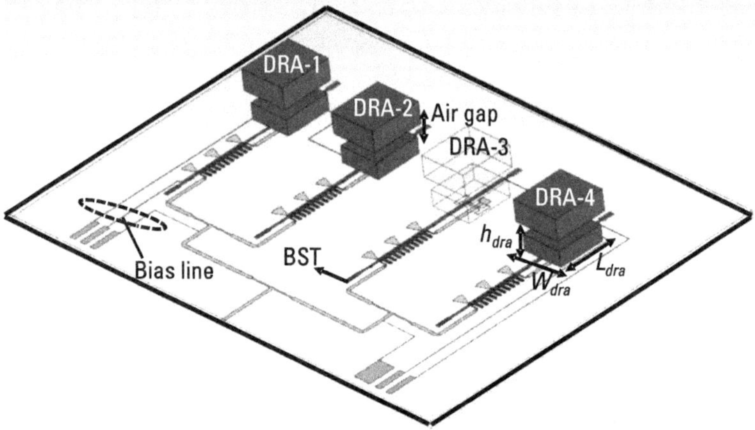

Figure 4.32 Layout of the four-element DRA with a feeding network proposed in [46].

A wideband series-fed linear DRA is explored in [47]. To improve the impedance bandwidth of the DRA, feeding rings under the DRA are incorporated, and slot windows are etched on the ground plane. The Chebyshev amplitude-distribution technique is used to achieve a low sidelobe level. Subsequently, slot windows with reflectors are used to enhance the overall performance of the DRA array. Figure 4.33 shows the structure of this series-fed linear DRA array. Eight-ceramic-cube DRA is fabricated of dielectric material with a permittivity of $\varepsilon_r = 8.9$, and the design parameters are length = 9.8 mm, width = 9.8 mm, and height = 9.1 mm. The ceramic cubes are mounted on a Rogers RT/duroid 5880 substrate with a relative permittivity $\varepsilon_r = 2.2$. Due to the

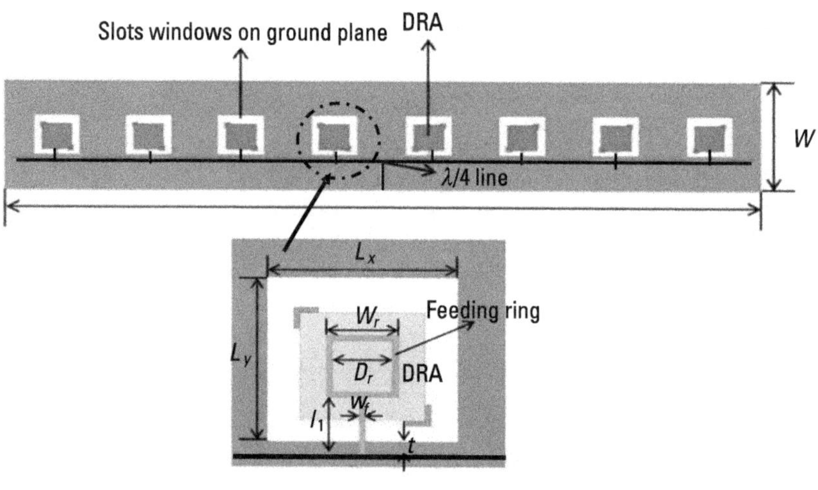

Figure 4.33 Structure of the series-fed linear DRA array proposed in [47].

optimized dimensions of the feeding ring, the resonant frequency of the feeding ring can be tuned with the DRA elements' resonant frequency, which can efficiently enhance the bandwidth of the DRA array. Slot windows are created on the ground plane to make the antenna array a bidirectional radiator; a reflector below the ground plane is incorporated to make the radiation unidirectional and to improve the directivity or gain of the antenna. The antenna offers a 6.57–9.08-GHz (31.9%) frequency band with a 15.7-dBi gain at 7.4 GHz.

In [48], some new techniques, such as the compound ground technique, are presented for the DRA and arrays. Also, the compound ground technique has good potential for K-band applications. A 2 × 2 antenna array achieves an 11-dBi gain. In [49], a 2 × 2 CDRA array with a microstrip aperture feed technique is successfully demonstrated to achieve a 13.9-dBi gain over the 6.62–7.15-GHz frequency range [49]. Each element of the CDRA array is constructed with $r = h = 10$ mm and relative permittivity $\varepsilon_r = 10$ dielectric material. The primary feed line is a rectangular aperture ($s_w \times s_l$) coupled through a microstrip line on a grounded substrate s_1. The DRA is isolated from the ground plane by introducing a thin dielectric layer s_2 to generate the $HEM_{12\delta}$ mode. In addition, to improve the antenna's performance, a metallic ring is integrated on top of the DRA. The metallic boundary forces the horizontal E fields to terminate on the ring surface, thereby maintaining the desired field configurations and impedance matching. Figure 4.34 shows this fabricated DRA array prototype.

Similarly, in [50], substrate-integrated waveguide-fed monolithic grid 4 × 4 and 8 × 8 DRA arrays obtained ($S_{11} < -10$ dB) impedance bandwidths of 10.4% and 12.0% with 15.2- and 19.4-dBi gains, respectively. An artificial grid DRA (GDRA) geometry integrated with a longitudinal slot-fed substrate integrated waveguide networks is designed at the V-band (60 GHz) for wideband and high-gain applications. Figure 4.35 shows the 4 × 4 and 8 × 8 monolithic GDRA array prototypes. The proposed antenna array is excited by SIW-based T-junction power splitters. All GDRA array layers are fabricated with X-ray

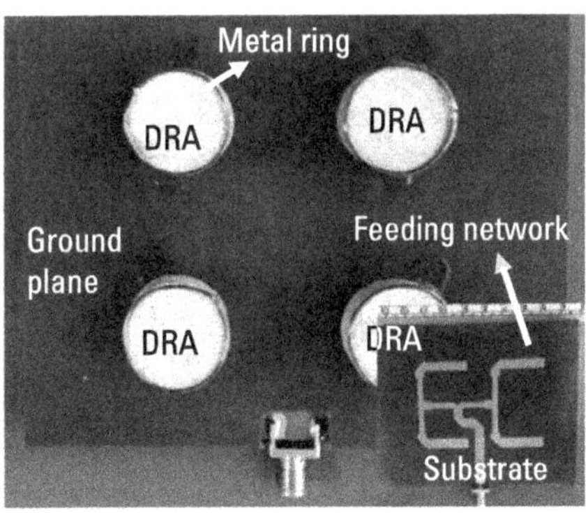

Figure 4.34 Fabricated DRA array prototype [49].

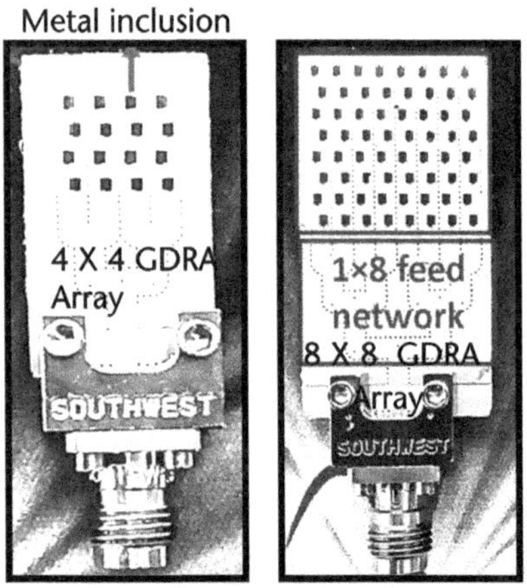

Figure 4.35 A 4 × 4 and 8 × 8 monolithic GDRA array prototype [50].

lithography and metal electroplating, whereas the feeding circuits are made of PCB technology on Rogers substrate with a relative permittivity of 2.95.

A novel aperture feeding structure with 2 × 2 wideband dual-polarized DRAs is explored for high gain in [51]. The proposed feeding mechanism allows a hybrid operation in slot and DRA mode; hence, broadband performance with gain is achieved. The feeding network is constructed of 50-Ω and 100-Ω microstrip lines that are 0.484 mm and 0.105 mm in width, respectively. For the feeding network, a T-junction with a 50-Ω line and a quarter-wave impedance transformer are used. Figure 4.36 shows the fabricated prototype of the antenna geometry and feeding network. For a 2 × 2 DRA array, a maximal gain of 10.4 dBi is obtained for port 1 at 30 GHz; the maximum gain for port 2 is 10.1 dBi at 36.8 GHz.

Another feeding technique, the dielectric image guide, has been used for exciting various antenna array systems. In [52], for example, a double-sided Taylor-distribution DRA array using a dielectric insular image guide (DIIG) is discussed for achieving high gain. To improve the gain, the dielectric insular

Figure 4.36 Fabricated prototype of the antenna geometry and feeding network proposed in [51].

resonator antenna array (DIRA) is constructed double-sided (i.e., a mirror image of the array is placed on the other side of the DIIG). Here, the Taylor distribution technique is used to suppress the sidelobe level. A 10-element linear DIRA array is fabricated and experimentally verified. It offers a 15.8-dBi high gain at a 36-GHz frequency. Figure 4.37 illustrates the geometry of the DIRA and the double-sided Taylor-distributed DIRA array prototype. Similarly, in [53], an aperture-coupled RDRA array fed by a DIIG is reported for high gain. A seven-element array and an 11-element array attain a 10.27-dBi and 13.9-dBi gain, respectively.

Next, in [54], a SIW-based four-element DRA configuration is reported for an 11.79-dBi gain. Further, in [55], a wideband high-gain RDRA array is demonstrated for millimeter-wave applications. The DRA size is enlarged by incorporating a high-pass filtering SIW cavity. The enlarged DRA dimension offers a higher DRA operating frequency without reducing the DRA size. The SIW cavity offers a larger fabrication tolerance to the DRA dimension. A 4 × 4 DRA array integrated with a backed cavity is designed for achieving a 16.4% impedance bandwidth with a 17.2-dBi gain. Figure 4.38 shows the antenna design configuration topology of each layer.

4.5 Enhancing Gain by Stacking Various Materials

This section considers stacking various dielectric materials in the construction of DRAs to enhance the gain of the antennas [56–61]. In DRAs, a stacked dielectric configuration offers various dielectric boundary conditions that enhance the E field or H field component intensity at the radiation boundary. Moreover, the aperture area and radiation boundary of the antenna are changed for the radiation, which leads to gain enhancement (Figure 4.39). Another ap-

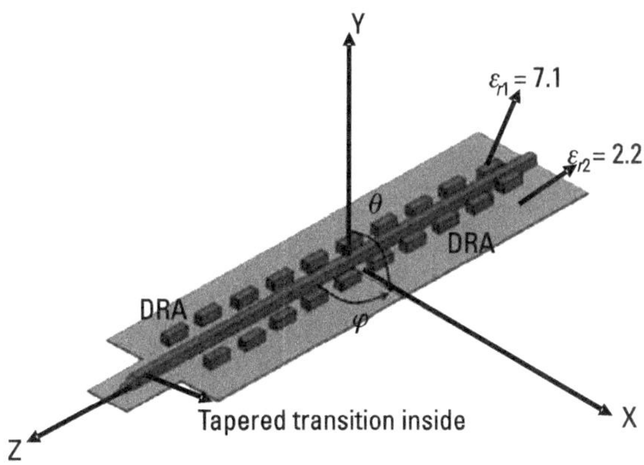

Figure 4.37 Geometry of a DIRA and a double-sided Taylor-distributed DIRA array [52].

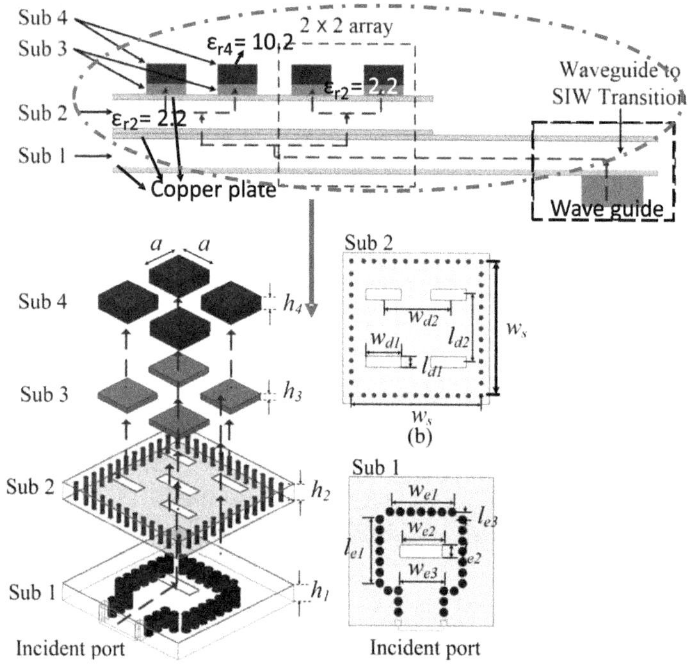

Figure 4.38 Antenna design configuration topology of each layer (4 × 4 array) [55].

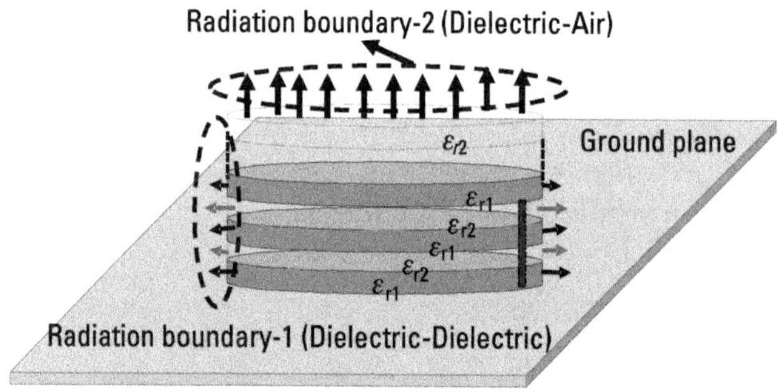

Figure 4.39 Stacking dielectric material DRA geometry with a radiation boundary.

proach in which alternate layers of two different anisotropic materials are arranged in different orientations is also distinguished for gain improvement.

A low-profile stacked DRA with a high gain and wide bandwidth is reported in [56]. For construction of the DRA, two dielectric layers with different permittivities ($\varepsilon_{r1} = 15$ and $\varepsilon_{r2} = 2.2$) are used. It has been observed that by properly selecting the permittivity and aspect ratio of the DRA dimensions, the frequency response and gain can be controlled. To make a low-profile antenna configuration, two different dielectric layers of the same length ($L >> h_1 + h_2$), but different heights of h_1 and h_2, are selected. The antenna is excited at the center of the microstrip-coupled slot, which is fabricated on a substrate with a permittivity of $\varepsilon_r = 3.38$. Figure 4.40(a) illustrates this antenna configuration. The antenna offers 40.4% (3.97–5.98-GHz) and 42.4% (3.94–6.06-GHz) simulated and measured impedance bandwidths, respectively. Two resonant modes are observed in the frequency band. Two radiating modes (TE^y_{111} and TE^y_{131}) are generated at the lower (4.35-GHz) and upper (5.4-GHz) frequencies. Two higher-order modes are excited due to different dielectric materials. A 10.5-dBi maximum gain is attained at 5.4 GHz, and the average gain across the frequency band is 9 dBi; the gain and impedance bandwidth plot are shown in Figure 4.40(b). The antenna can help achieve desired requirements, such as a low profile, a wide bandwidth, and a high gain, by properly selecting the aspect ratio (length-to-height) and the permittivity of the two dielectric layers.

Further, in [57], uniaxial anisotropic materials are used in RDRAs to enhance the radiation from their side walls, which leads to the improvement of boresight gain. The interaction of EM waves with different boundary conditions

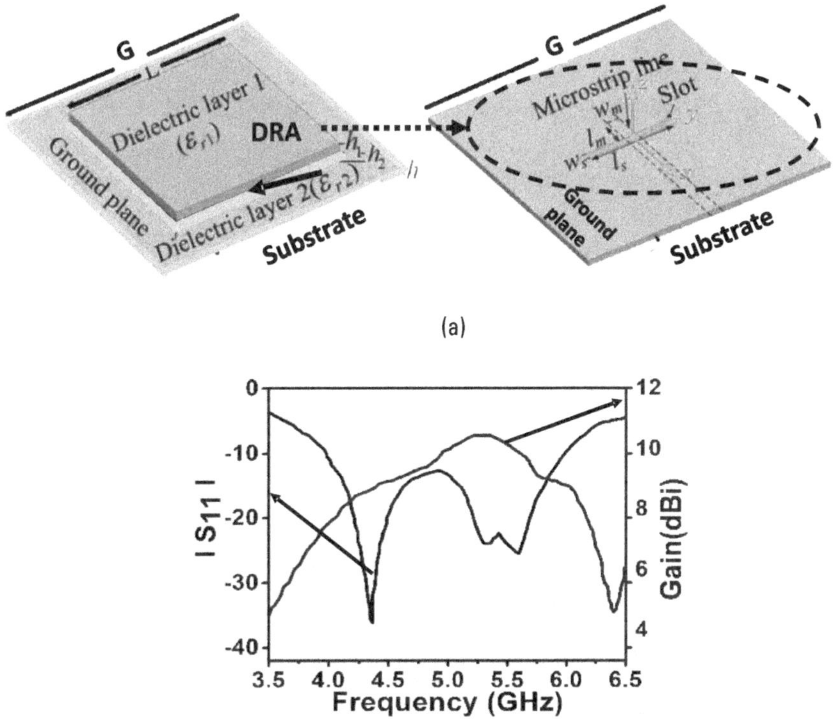

Figure 4.40 Antenna configuration: (a) the stacking of two dielectric layers and (b) the impedance bandwidth and gain [56].

on the DRA walls offers an impedance bandwidth of 20.65% (3.17–3.9-GHz) and a peak broadside gain of 8.4 dBi.

The stacked dielectric sheet structure behaves like a homogeneous uniaxial anisotropic dielectric when the thickness of the sheets is small compared to the wavelength (less than one-tenth of a wavelength). For the fabrication of an RDRA, 14 layers [seven layers of TMM13i ($\varepsilon_1 = 12.2$) and seven layers of Duroid 5880 ($\varepsilon_2 = 2.2$)] are stacked together mechanically using a jig and fixture. Each sheet of thickness $d_1 = 1.25$ mm and a dielectric constant ε_1 are separated by a sheet of thickness $d_2 = 1.575$ mm and dielectric constant ε_2, and this antenna geometry is excited by aperture microstrip feeding techniques, shown in Figure 4.41. The total effective permittivity tensor can be calculated with (4.20). Also, whether the electromagnetic wave propagates a signal in a uniaxial anisotropic medium depends on permittivity tensor of uniaxial anisotropic material.

$$\varepsilon_x = \varepsilon_y = \frac{\varepsilon_1\varepsilon_2\left(d_1 + d_2\right)}{\varepsilon_1 d_1 + \varepsilon_2 d_2}, \varepsilon_z = \frac{\varepsilon_1 d_1 + \varepsilon_2 d_2}{d_1 + d_2} \tag{4.20}$$

Similarly, in [58], a compact uniaxial anisotropic rectangular DRA is reported for high gain. This proposed antenna consists of a multilayer uniaxial anisotropy dielectric structure and is excited by a probe feed that acts as a PEC symmetry plane. Also, a higher-order radiating mode is excited, which offers high gain. This antenna configuration offers a 19.4% wide impedance bandwidth with a 8.1-dB maximum gain.

In [59], cylindrical anisotropic DRA is investigated for high gain. To construct the proposed CDRA, uniaxial anisotropic material is obtained by stacking parallel layers of two different dielectrics. The antenna anisotropy radiation boundary of the material enhances the gain in the boresight direction. Antenna gain can also be improved by selecting optimized design parameters (radius-to-height ratio). Figure 4.42 shows the geometry of this proposed antenna [59] with excitation techniques, which attained an impedance bandwidth of 16.6%

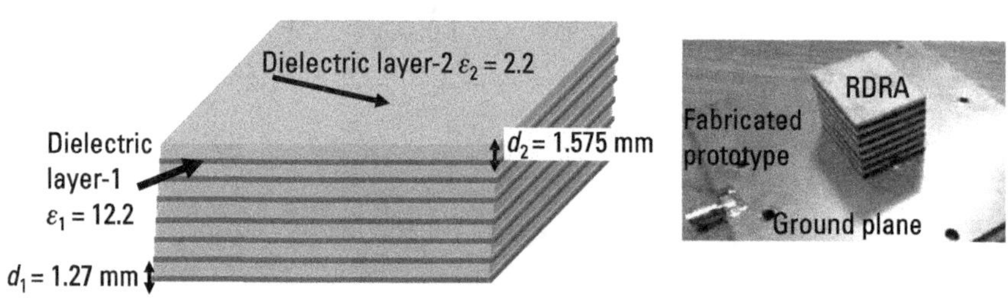

Figure 4.41 Multilayer-structure RDRA: (a) stacking uniaxial medium and (b) fabricated RDRA [57].

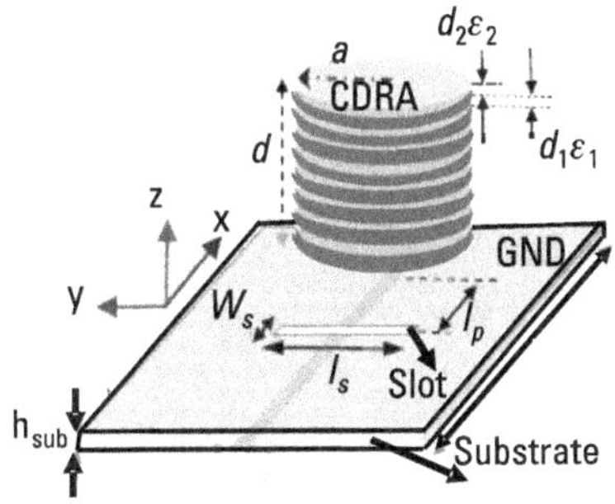

Figure 4.42 Geometry of the antenna with excitation techniques proposed in [59].

(3.15–3.72-GHz) and a peak gain of 8.4 dBi; this is useful for various wireless applications.

In [60], the author discusses the use of equilateral triangular uniaxial-anisotropic DRAs to increase gain in the boresight direction and impedance bandwidth. Gain improvement occurs due to an increase in the sidewall radiations, and bandwidth enhancement occurs when the effective permittivity of the structure is reduced. This antenna offers a 27.7% (3.025–4-GHz) impedance bandwidth and a more than 8-dBi gain for the complete frequency range.

Another method of engraving grooves on the side walls of an RDRA is investigated in [61] for attaining a 9.6-dBi gain. By incorporating corrugated engraving grooves on the side walls, the radiations from the side wall boundary of the DRA increase, thereby improving which gain. The fabricated prototype consists of 10 layers of a substrate $\varepsilon_r = 10.2$, with a thickness of 1.575 mm for each layer. An excitation microstrip line is printed on a substrate ($\varepsilon_r = 3.38$) with a thickness of 0.508 mm. Figure 4.43 presents this fabricated prototype and its design configuration. Similarly, in [62], an RDRA with corrugated walls achieved a 9-dBi gain.

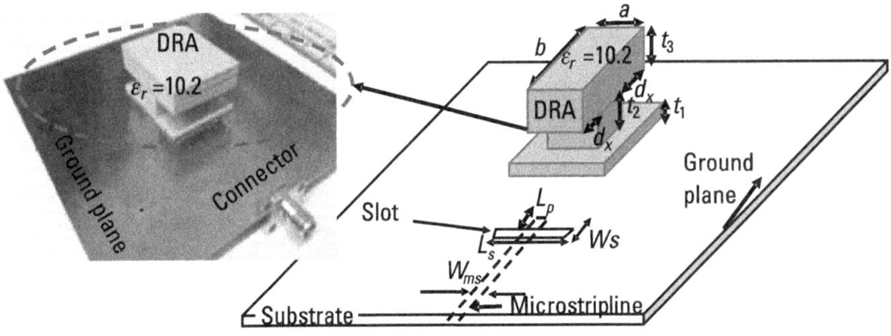

Figure 4.43 Fabricated prototype and its design configuration [61].

4.6 Conclusion

This chapter discussed various techniques for the improvement of gain in DRA, including some of the fundamental theory of the antenna radiation mechanism, aperture area, and perturbation theory to explain the radiation phenomena. The chapter explored several new techniques for achieving a high gain with a wide bandwidth, including metamaterial top loading, reflector surface, and integrating metal to surround the DRA, using higher-order modes, exciting arrays of DRAs with different feeding techniques, and stacking uniaxial material. All the salient features of these techniques were briefly explained. This chapter also covered DRA fabrication techniques and the design processes.

References

[1] Balanis C. A., *Antenna Theory Analysis Design,* John Wiley & Sons, 2005.

[2] Harington, R. F., *Time Harmonic Electromagnetic Fields,* New York: IEEE Press, 2001.

[3] Waldron, R. A., "Perturbation theory of resonant cavities," *Proc. IEE Part C: Monographs,* Vol. 107, pp. 272–274, 1960.

[4] Sahu, B., et al., "Dual Segment Rectangular Dielectric Resonator Antenna with Metamaterial for Improvement of Bandwidth and Gain," *International Journal of RF and Microwave Computer-Aided Engineering,* Vol. 24, No. 6, 2014, pp. 646–655.

[5] Sinha, M., V. Killamsetty, and B. Mukherjee, "Near Field Analysis of RDRA Loaded with Split Ring Resonators Superstrate," *Microwave and Optical Technology Letters,* Wiley, Vol. 60, Issue 2, 2018, pp. 472–478.

[6] Gupta, R. D., and M. S. Parihar, "Differentially Driven Wideband Fabry–Perot Cavity Antenna," *IET Microwaves, Antennas & Propagation,* Vol. 13, No. 13, 2019, pp. 2365-2371.

[7] Akbari, M., et al., "Gain Enhancement of Circularly Polarized Dielectric Resonator Antenna Based on FSS Superstrate for MMW Applications," *IEEE Transactions on Antennas and Propagation,* Vol. 64, No. 12, Dec. 2016, pp. 5542–5546.

[8] Panda, A. K., S. Sahu, and R. K. Mishra, "DRA Gain Enhancement Using a Planar Metamaterial Superstrate," *Int J RF Microw Comput Aid Eng.*, Vol. 28, No. 7, 2018, pp. 1–10.

[9] Pandey, A. K., et al., "High Gain Compact Rectangular Dielectric Resonator Antenna Using Metamaterial as Superstrate," *International Journal of RF and Microwave Computer-Aided Engineering,* Wiley, Vol. 29, Issue 12, 2019, pp. 1–10.

[10] Chauhan, M., A. Rajput, and B. Mukherjee,"Wideband Circularly Polarized Low Profile Dielectric Resonator Antenna with Meta Superstrate for High Gain," *AEU—International Journal of Electronics and Communications,* Volume 128, 2021.

[11] Kumar, A., et al., "Metamaterial Loaded Aperture Coupled Biodegradable Star-Shaped Dielectric Resonator Antenna for WLAN and Broadband Applications," *Microwave and Optical Technology Letters,* 2019.

[12] Dutta, K., et al., "New Approach in Designing Resonance Cavity High-Gain Antenna Using Nontransparent Conducting Sheet as the Superstrate," *IEEE Transactions on Antennas and Propagation,* Vol. 63, No. 6, June 2015, pp. 2807–2813.

[13] Dash, S. K. K., et al., "Gain Improvement of Cylindrical Dielectric Resonator Antenna Using Flat Reflector Plane: A New Approach," *IET Microwaves, Antennas & Propagation,* Vol. 11, No. 11, 2017, pp. 1622–1628.

[14] Agrawal, S., et al., "A Wideband High Gain Dielectric Resonator Antenna for RF Energy Harvesting Application," *AEU—International Journal of Electronics and Communications,* Vol. 78, 2017, pp. 24–31.

[15] Dash, S. K. K., and T. Khan, "Wideband High Gain Conical Dielectric Resonator Antenna: An Experimental Study of Superstrate and Reflector," *International Journal of RF and Microwave Computer-Aided Engineering,* Vol. 27, No. 9, 2017.

[16] Wen, J., et al., "Wideband Circularly Polarized Dielectric Resonator Antenna Loaded with Partially Reflective Surface," *International Journal of RF and Microwave Computer-Aided Engineering.* 2019.

[17] Denidni, T. A., Y. Coulibaly, and H. Boutayeb, "Hybrid Dielectric Resonator Antenna with Circular Mushroom-Like Structure for Gain Improvement," *IEEE Transactions on Antennas and Propagation,* Vol. 57, No. 4, April 2009, pp. 1043–1049.

[18] Perron, A., T. A. Denidni, and A. Sebak, "High-Gain Hybrid Dielectric Resonator Antenna for Millimeter-Wave Applications: Design and Implementation," *IEEE Transactions on Antennas and Propagation,* Vol. 57, No. 10, Oct. 2009, pp. 2882–2892.

[19] Abdel Wahab, W. M., D. Busuioc, and S. Safavi–Naeini, "Low Cost Planar Waveguide Technology-Based Dielectric Resonator Antenna (DRA) for Millimeter-Wave Applications: Analysis, Design, and Fabrication," *IEEE Transactions on Antennas and Propagation,* Vol. 58, No. 8, Aug. 2010, pp. 2499–2507.

[20] Tong, C., et al., "Differentially Coplanar-Fed Filtering Dielectric Resonator Antenna for Millimeter-Wave Applications," *IEEE Antennas and Wireless Propagation Letters,* Vol. 18, No. 4, April 2019, pp. 786–790.

[21] Hu, P. F., et al., "A Compact Filtering Dielectric Resonator Antenna with Wide Bandwidth and High Gain," *IEEE Transactions on Antennas and Propagation,* Vol. 64, No. 8, Aug. 2016, pp. 3645–3651.

[22] Cicchetti, R., et al., "A High–Gain Mushroom-Shaped Dielectric Resonator Antenna for Wideband Wireless Applications," *IEEE Transactions on Antennas and Propagation,* Vol. 64, No. 7, July 2016, pp. 2848–2861.

[23] Chang, T., and J. Kiang, "Sectorial-Beam Dielectric Resonator Antenna for WiMAX with Bent Ground Plane," *IEEE Transactions on Antennas and Propagation,* Vol. 57, No. 2, Feb. 2009, pp. 563–567.

[24] Nasimuddin, and K. P. Esselle, "A Low-Profile Compact Microwave Antenna with High Gain and Wide Bandwidth," *IEEE Transactions on Antennas and Propagation,* Vol. 55, No. 6, June 2007, pp. 1880–1883.

[25] Gupta, R. D., and M. S. Parihar, "Differentially Fed Wideband Rectangular DRA with High Gain Using Short Horn," *IEEE Antennas and Wireless Propagation Letters,* Vol. 16, 2017, pp. 1804–1807.

[26] Chauhan, M., and B. Mukherjee, "High Gain Cylindrical Dielectric Resonator Integrated with Horn for Multiband Wireless Applications," *International Journal of RF and Microwave Computer-Aided Engineering*, Wiley, Vol. 30, Issue 5, 2020, pp. 1–9.

[27] Nasimuddin, and K. P. Esselle, "Antennas with Dielectric Resonators and Surface Mounted Short Horns for High Gain and Large Bandwidth," *IET Microwaves, Antennas & Propagation*, Vol. 1, No. 3, June 2007, pp. 723–728.

[28] Wang, Y., et al., "Design of High Gain, Broadband Cylindrical Dielectric Resonator Antenna," *Electronics Letters*, Vol. 49, No. 24, 21 November 2013, pp. 1506–1507.

[29] Feng, K., et al., "Study on Dielectric Resonator Antenna with Annular Patch for High Gain and Large Bandwidth," *Chinese Journal of Electronics*, Vol. 24, No. 4, 2015, pp. 869–872.

[30] Chauhan, M., A. K. Pandey, and B. Mukherjee, "A Novel Compact Cylindrical Dielectric Resonator Antenna for Wireless Sensor Network Application," *IEEE Sensors Letters*, Vol. 2, No. 2, June 2018, pp. 1–4.

[31] Kim, T., and S. Pak, "Enhanced Gain and Miniaturization Method of Stacked Dielectric Resonator Antenna Using Metallic Cap," *IET Microwaves, Antennas & Propagation*, Vol. 13, No. 8, July 2019, pp. 1198–1201.

[32] Petosa, A., *Dielectric Resonator Antennas Handbook*, Norwood, MA: Artech House, 2007.

[33] Petosa, A., S. Thirakoune, and A. Ittipiboon, "Higher-Order Modes in Rectangular DRAs for Gain Enhancement," *2009 13th International Symposium on Antenna Technology and Applied Electromagnetics and the Canadian Radio Science Meeting*, Banff, AB, Canada, 2009, pp. 1–4.

[34] Petosa, A., and S. Thirakoune, "Rectangular Dielectric Resonator Antennas with Enhanced Gain," *IEEE Transactions on Antennas and Propagation*, Vol. 59, No. 4, April 2011, pp. 1385–1389.

[35] Pan, Y., K. W. Leung, and K. Luk, "Design of the Millimeter-Wave Rectangular Dielectric Resonator Antenna Using a Higher-Order Mode," *IEEE Transactions on Antennas and Propagation*, Vol. 59, No. 8, Aug. 2011, pp. 2780–2788.

[36] Li, C., and T. Chiu, "340-GHz Low-Cost and High-Gain On-Chip Higher Order Mode Dielectric Resonator Antenna for THz Applications," *IEEE Transactions on Terahertz Science and Technology*, Vol. 7, No. 3, May 2017, pp. 284–294.

[37] Abdulmajid, A. A., Y. Khalil, and S. Khamas, "Higher-Order-Mode Circularly Polarized Two-Layer Rectangular Dielectric Resonator Antenna," *IEEE Antennas and Wireless Propagation Letters*, Vol. 17, No. 6, June 2018, pp. 1114–1117.

[38] Guha, D., et al., "Higher Order Mode Excitation for High-Gain Broadside Radiation from Cylindrical Dielectric Resonator Antennas," *IEEE Transactions on Antennas and Propagation*, Vol. 60, No. 1, Jan. 2012, pp. 71–77.

[39] Guha, D., P. Gupta, and C. Kumar, "Dualband Cylindrical Dielectric Resonator Antenna Employing $HEM_{11\delta}$ and $HEM_{12\delta}$ Modes Excited by New Composite Aperture," *IEEE Transactions on Antennas and Propagation*, Vol. 63, No. 1, Jan. 2015, pp. 433–438.

[40] Mrnka, M., and Z. Raida, "Enhanced-Gain Dielectric Resonator Antenna Based on the Combination of Higher-Order Modes," *IEEE Antennas and Wireless Propagation Letters,* Vol. 15, 2016, pp. 710–713.

[41] Sharma, A., et al., "Novel Feeding Mechanism To Stimulate Triple Radiating Modes in Cylindrical Dielectric Resonator Antenna," *IEEE Access,* Vol. 4, 2016, pp. 9987–9992.

[42] Yang, L., et al., "High-Gain Magnetic-Dipole Quasi-Yagi Antenna Using Higher-Order-Mode Dielectric Resonator with Near-Zero-Index Metamaterial," *International Journal of RF and Microwave Computer-Aided Engineering,* 2020.

[43] Bit-Babik, G., C. DiNallo, and A. Faraone, "Multimode Dielectric Resonator Antenna of Very High Permittivity," *IEEE Antennas and Propagation Symposium 2004,* Vol. 2, Monterey, CA, June 2004, pp. 1383–1386.

[44] Li, B., and K. W. Leung, "A Wideband Strip-Fed Rectangular Dielectric Resonator Antenna," *IEEE Antennas and Propagation Symposium 2005,* Vol. 2, Washington, D.C., July 2005, pp. 172–175.

[45] Ranjbar Nikkhah, M., J. Rashed-Mohassel, and A. A. Kishk, "High-Gain Aperture Coupled Rectangular Dielectric Resonator Antenna Array Using Parasitic Elements," *IEEE Transactions on Antennas and Propagation,* Vol. 61, No. 7, July 2013, pp. 3905–39081.

[46] Nikfalazar, M., et al., "Steerable Dielectric Resonator Phased-Array Antenna Based on Inkjet-Printed Tunable Phase Shifter with BST Metal-Insulator-Metal Varactors," *IEEE Antennas and Wireless Propagation Letters,* Vol. 15, 2016, pp. 877–880.

[47] Lin, J., W. Shen, and K. Yang, "A Low-Sidelobe and Wideband Series-Fed Linear Dielectric Resonator Antenna Array," *IEEE Antennas and Wireless Propagation Letters,* Vol. 16, 2017, pp. 513–516.

[48] Sarkar, C., D. Guha, and C. Kumar, "Glueless Compound Ground Technique for Dielectric Resonator Antenna and Arrays," *IEEE Antennas and Wireless Propagation Letters,* Vol. 16, 2017, pp. 2440–2443.

[49] Sarkar, C., D. Guha, and C. Kumar, "Advanced Design of Higher Mode-Based Dielectric Resonator Antenna Array Featuring High-Gain Operation Over a Large Frequency Band," *Microwave and Optical Technology Letters,* 2020.

[50] Mazhar, W., et al., "60 GHz Substrate Integrated Waveguide-Fed Monolithic Grid Dielectric Resonator Antenna Arrays," *IEEE Antennas and Wireless Propagation Letters,* Vol. 18, No. 6, June 2019, pp. 1109–1113.

[51] Kowalewski, J., et al., "A Millimeter-Wave Broadband Dual-Polarized Dielectric Resonator Antenna Based on Hybrid Modes," *IEEE Antennas and Wireless Propagation Letters,* Vol. 19, No. 7, July 2020, pp. 1068–1072.

[52] Jin, L., R. Lee, and I. Robertson, "A Dielectric Resonator Antenna Array Using Dielectric Insular Image Guide," *IEEE Transactions on Antennas and Propagation,* Vol. 63, No. 2, Feb. 2015, pp. 859–862.

[53] Al-Zoubi, A. S., A. A. Kishk, and A. W. Glisson, "Aperture Coupled Rectangular Dielectric Resonator Antenna Array Fed by Dielectric Image Guide," *IEEE Transactions on Antennas and Propagation,* Vol. 57, No. 8, Aug. 2009, pp. 2252–2259.

[54] Abdallah, M., et al., "A Tunable Circuit Model for the Modeling of Dielectric Resonator Antenna Array," *IEEE Antennas and Wireless Propagation Letters,* Vol. 15, 2016, pp. 830–833.

[55] Chen, Z., et al., "Millimeter-Wave Rectangular Dielectric Resonator Antenna Array With Enlarged DRA Dimensions, Wideband Capability, and High-Gain Performance," *IEEE Transactions on Antennas and Propagation,* Vol. 68, No. 4, April 2020, pp. 3271–3276.

[56] Pan, Y. M., and S. Y. Zheng, "A Low-Profile Stacked Dielectric Resonator Antenna with High-Gain and Wide Bandwidth," *IEEE Antennas and Wireless Propagation Letters,* Vol. 15, 2016, pp. 68–71.

[57] Fakhte, S., H. Oraizi, and L. Matekovits, "High Gain Rectangular Dielectric Resonator Antenna Using Uniaxial Material at Fundamental Mode," *IEEE Transactions on Antennas and Propagation,* Vol. 65, No. 1, Jan. 2017, pp. 342–347.

[58] Fakhte, S., and H. Oraizi, "Compact Uniaxial Anisotropic Dielectric Resonator Antenna Operating at Higher Order Radiating Mode," *Electronics Letters,* Vol. 52, No. 19, 2016, pp. 1579–1580.

[59] Fakhte, S., et al., "Cylindrical Anisotropic Dielectric Resonator Antenna with Improved Gain," *IEEE Transactions on Antennas and Propagation,* Vol. 65, No. 3, March 2017, pp. 1404–1409.

[60] Fakhte, S., I. Aryanian, and L. Matekovits, "Analysis and Experiment of Equilateral Triangular Uniaxial-Anisotropic Dielectric Resonator Antennas," *IEEE Access,* Vol. 6, 2018, pp. 63071-63079.

[61] Fakhte, S., H. Oraizi, and L. Matekovits, "Gain Improvement of Rectangular Dielectric Resonator Antenna by Engraving Grooves on Its Side Walls," *IEEE Antennas and Wireless Propagation Letters,* Vol. 16, 2017, pp. 2167–2170.

[62] Fakhte, S., L. Matekovits, and I. Aryanian, "Rectangular Dielectric Resonator Antenna with Corrugated Walls," *IEEE Access,* Vol. 7, 2019, pp. 3422–3429.

5

Multiple-Input and Multiple-Output DRAs and Diversity Applications

The necessity for high data rates and fast communication systems has escalated the overall system requirements for futuristic communication applications. In these challenges, it becomes mandatory to operate on multiple-input and multiple-output (MIMO)-based applications rather than single-feed antennas. Further accelerating this trend is the advancement toward fifth- and sixth-generation mobile communications; since all devices are likely to become smart and mobile-controlled using the Internet of Things (IoT), MIMO antennas become a prudent solution.

It is interesting to note that for MIMO antennas, the isolation between the individual antennas is very important. Conventional metallic antennas offer a boundary condition where the tangential E field components cease, while the charge distribution of dielectric boundaries connected in a multiple pattern creates a discontinuity at the surface. This charge distribution may give rise to coupling between the DRA elements, and as a result, achieving suitably high isolation between DRA elements is a major challenge in MIMO DRA designs.

MIMO systems were created in response to the limitations of single-input, single-output (SISO) systems. The gigabit-per-second data rates of SISOs require a very large frequency spectrum that makes them unable to cooperate in non-LOS (NLOS) networks. Also, SISOs require a very large SNR in practical receivers. Achieving an average signal-to–interference with noise ratio (SINR) > 10 dB is critical in a SISO, even with spectral efficiencies ≥ 4 bits/s/Hz. Moreover, due to power constraints, high-gain SISOs result in scattering, and thus, they are unable to assist in non-LOS (NLOS) communication. Extending the bandwidth in SISOs to achieve a gigabit-per-second data rate limits their range and reduces their fade margin. Hence, increased power is required for wide coverage. A SISO reduces range by three times in comparison with modern wireless devices [1].

In a MIMO, each spatial signal is transmitted from a separate antenna chain in the same frequency channel as the transmitter. The receiver receives each signal on each of its identical antenna chains. Since the receiver knows the phase offsets of its own antennas, it can reconstruct the original signal. A MIMO allows for more uniformly distributed energy across the given coverage area and devices and therefore aids in improving coverage issues as compared to the conventional SISO antennas. Essentially a MIMO employs multiple antennas on the receiver and transmitter to utilize the multipath effects that always exist to transmit additional data, rather than causing interference. Current research, however, has investigated MIMO DRAs with a single-DRA structure with multiple feed lines as well. The multifeed setup helps to reduce the size of the overall MIMO antennas.

With the increase it provides in throughput and SNR, MIMO is one of the key enabling techniques for fifth-generation wireless technology. The technologies that MIMO antennas facilitate include the IoT, hyperdense cellularization for network planning, connecting multiple devices for a single user, smart devices, and remote medical monitoring. Accordingly, many researchers have focused on microstrip patch-based MIMO antennas for enhancing channel capacity.

5.1 MIMO Antenna Characteristics and Attributes

Before discussing MIMO DRA geometries and the techniques used to achieve them, it is important to review some important associated components. Sections 5.1.1–5.1.6 cover these terms.

5.1.1 Reflection Coefficient (R)

The reflection coefficient is a parameter that describes how much of a wave is reflected by an impedance discontinuity in the transmission medium. For an antenna, it describes how much power is reflected back because of impedance mismatch. In terms of the forward and reflected waves determined by the voltage and current, the reflection coefficient is defined as the complex ratio of the voltage of the reflected wave (V^-) to that of the incident wave (V^+). This is shown in (5.1). In addition, note that the reflection coefficient is based on load termination (Z_L) and the characteristic impedance of the transmission line (Z_0). This is also given in (5.1).

$$\Gamma = \frac{V^-}{V^+} = \frac{Z_L - Z_0}{Z_L + Z_0} \quad (5.1)$$

In terms of S parameters, the return loss (RL) for a one-port device (antenna) is given as

$$RL = -20\log_{10}|S_{11}|dB \quad (5.2)$$

If RL is 0 dB, then it means that the total power is reflected back and that no power is radiated by the antenna. For measuring the impedance bandwidth, an RL of 10 dB is considered. Also, corresponding to the 10-dB line, a VSWR of less than 2 is obtained.

5.1.2 Transmission Coefficient (τ)

A transmission coefficient describes the amplitude, intensity, or total power of a transmitted wave relative to an incident wave. The transmission coefficient is the ratio of the amplitude of the complex transmitted wave to that of the incident wave at a discontinuity in the transmission line. Conventionally, the transmission coefficient (τ) and the reflection coefficient are given as

$$1 + \Gamma = \tau \quad (5.3)$$

5.1.3 Envelope Correlation Coefficient

The envelope correlation coefficient (ECC) suggests how much of two antennas' radiation patterns are independent. If one antenna is completely horizontally polarized (H-polarized), and the other is completely vertically polarized (V-polarized), then the two antennas are said to have a correlation of zero. Similarly, if one antenna only radiated energy toward the boreside, and the other only radiated energy toward the ground, these antennas would also have an ECC of zero. Hence, the ECC takes into account the antennas' radiation pattern shape, polarization, and even the relative phase of the fields between the two antennas. It shows the correlation between different exciting ports, and the diversity gain (DG) determines the improvement in the SNR of multiple exciting ports to a single exciting port from the receiver antenna.

The ECC is calculated [2] by

$$\text{ECC} = \frac{\left|s_{11}^{*}\ s_{12} + s_{21}^{*}\ s_{22}\right|}{\left(1 - |s_{11}|^2 - |s_{21}|^2\right)\left(1 - |s_{22}|^2 - |s_{12}|^2\right)} \quad (5.4)$$

where $|S_{11}|$, $|S_{22}|$ represents the reflection coefficient parameters of port 1 and port 2. $|S_{12}|$, $|S_{21}|$ represents the isolation between port 1 and port 2, respectively. Also DG depends on ECC. To calculate the DG of a proposed antenna, (5.5) is followed [2].

$$\mathrm{DG} = \sqrt{1 - \mathrm{ECC}} \tag{5.5}$$

5.1.4 Channel Capacity Loss

Channel capacity loss (CCL) is another important key element of a MIMO antenna system. For MIMO antenna performance, the CCL is calculated to determine the maximum attainable message rate at which a signal can be transmitted without loss of signal strength [2]. To calculate a proposed antenna's CCL, use (5.6).

$$\mathrm{CCL} = -\log_2 \det\left(\Psi^R\right) \tag{5.6a}$$

where Ψ^R is the receiving antenna port's correlation matrix, which is

$$\Psi^R = \begin{pmatrix} \beta_{11} & \beta_{12} \\ \beta_{21} & \beta_{22} \end{pmatrix} \tag{5.6b}$$

$$\beta_{11} = 1 - \left(|S_{11}|^2 + |S_{12}|^2\right),\ \beta_{12} = -\left(S_{11}^* S_{12} + S_{21}^* S_{22}\right)$$

$$\beta_{21} = -\left(S_{22}^* S_{21} + S_{12}^* S_{11}\right), \beta_{22} = 1 - \left(|S_{22}|^2 + |S_{21}|^2\right)$$

A CCL of less than 0.5 bps/Hz is acceptable for MIMO antenna system performance.

5.1.5 Mean Effective Gain

In a multipath environment, the ability of an antenna to receive EM power is quantified by the mean effective gain (MEG). It is defined as the ratio of the mean received power to the mean incident power of the antenna [3]. This is given as

$$\mathrm{MEG} = \oint \left[\frac{\mathrm{XPD}}{1 + XPD} p_\theta(\Omega) G_\theta(\Omega) + \frac{1}{1 + \mathrm{XPD}} p_v(\Omega) G_v(\Omega)\right] d\Omega \tag{5.7}$$

where XPD is the cross-polarization discrimination of the incident field (the ratio between the vertical and horizontal power densities), and p_θ and p_v are the θ and v components of the probability distribution functions of the incoming wave, respectively. G_θ and G_v are the power gain patterns of the antenna [3].

5.1.6 Total Active-Reflection Coefficient

The total active-reflection coefficient (TARC) relates the total incident power to the total outgoing power in an N-port microwave component. The TARC is mainly used for MIMO antenna systems and array antennas, where the outgoing power is unwanted reflected power. TARC is a function of frequency, and it also depends on the scan angle and tapering.

TARC is defined as the ratio of reflected power to the incident power of the exciting ports of a MIMO antenna system. With (5.8), TARC can be calculated from the S-parameters [2].

$$\mathrm{TARC} = \frac{\sqrt{\left(\left|S_{11} + S_{12}e^{j\theta}\right|^2\right) + \left|S_{21} + S_{22}e^{j\theta}\right|^2}}{\sqrt{2}} \tag{5.8}$$

5.2 Single-DRA Element with Dual- or Multiple-Feeding Techniques for MIMO Applications

One of the methods to deploy DRAs for MIMO applications is by using the degeneracy of the EM modes. This concept is used in [4] for LTE Femtocell Base Station applications. The mode degeneracy is achieved by perturbing the boundary of the DRA, which causes the amount of energy stored by a specific mode to change as well as the resonant frequency of that mode. Thus by introducing an appropriate boundary perturbation, the TE and HE modes of the DRA resonate at the same frequency and share the same impedance bandwidth. The perturbation is applied on the half-split CDRA where the TE mode can be excited. Further, the ports are introduced on the top curved surface and the lateral surface resting on the ground plane leading to a 2 × 2 MIMO DRA application. The proposed MIMO DRA was able to achieve a channel capacity of 11.1 b/s/Hz (as compared to a theoretical maximum 2 × 2 capacity of 13.4 b/s/Hz). This is shown in Figure 5.1.

The same half-split CDRA geometry frequency-agile MIMO DRA is explored in [5]. The frequency agility concept is achieved by appending reconfigurable parasitic slots in proximity to the dielectric resonator's radiating element. Because of the orthogonality between the two radiating ports, the operating

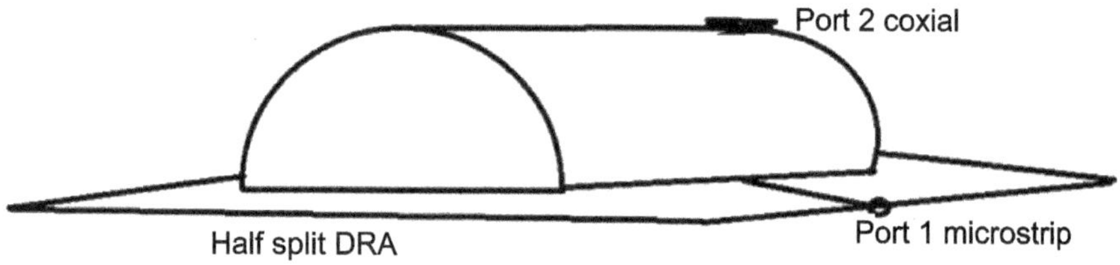

Figure 5.1 Half-split DRA for MIMO applications [4].

frequency of each port can be individually tuned without affecting the frequency response of the other port. Thus, the DRA operates on two modes of operation: multifrequency mode and MIMO mode. In the multifrequency mode, the antenna ports are independently swept through the frequency band of interest and look for nearby potential users.

Similarly, the antenna ports can be reconfigured into MIMO applications to boost the capacity whenever heavy data traffic is required and the channel condition is appropriate. These capabilities make the antenna suitable for dynamic spectrum-access applications in cognitive radios.

In [6], an RDRA for MIMO applications in 4G and LTE frequencies is explored. The RDRA is excited by a coaxial port with a shorting post and conformal strip in one end, and the other perpendicular port is excited by a coplanar waveguide (CPW) feed. This is as shown in Figure 5.2. The measured impedance bandwidth for port 1 and port 2 is 47% (2.09–3.38 GHz) and 25% (2.40–3.09 GHz), respectively. The measured correlation coefficient is 0.03 with a nearly 10-dB diversity gain at frequency 2.6 GHz. The MIMO RDRA yields an isolation of above 20 dB over the operating frequency. A gain of 4.97 dBi is obtained for port 1, and 4.51 dBi is obtained for port 2 at 2.6 GHz. The RDRA is fabricated from a permittivity of 4.4 (i.e., FR4).

In order to improve the diversity gain of a DRA, FSS and slots are deployed around the DRA as discussed in [7]. An FSS wall is inserted between the DRAs to reduce the free-space radiation. Then, two slots with different sizes acting together like an LC resonator are etched from the common ground plane

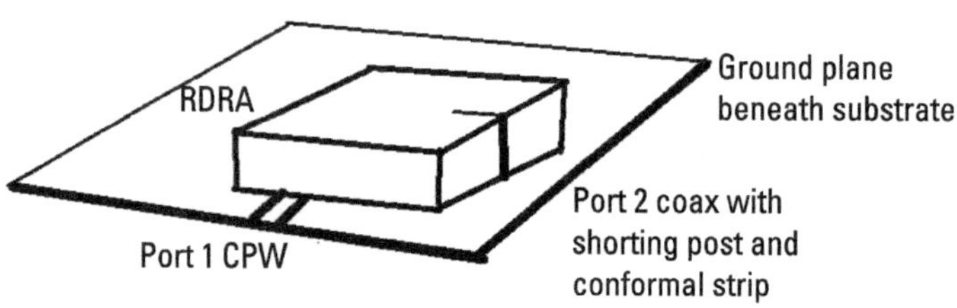

Figure 5.2 RDRA for MIMO application [6].

of the structure to reduce the surface current. The combined effect, explored for the millimeter-wave frequency at 60 GHz, aids in improving the directive gain and isolation between the DRAs. Similarly, in [8], three decoupled ports and near-degenerate modes are used to excite a block of RDRAs to achieve MIMO applications. In [8], three modes are excited, but to achieve decoupling two modes are excited such that there is low mutual spatial overlapping between their field magnitudes; then, a third mode is imposed such that its field components are perpendicular to the other two modes. Interestingly at the three ports, the impedance bandwidths offered are 880 MHz, 870 MHz, and 2.61 GHz, respectively. The wide impedance bandwidth of the third port is attributed to the overlap of the two resonances of the $TE^{y}_{2\delta 1}$ mode. The proposed antenna could be used either as a three-port MIMO antenna with an operating bandwidth of 720 MHz or as a two-port MIMO antenna with an operating bandwidth of 880 MHz at a center frequency of 9.48 GHz—achieving a high gain close to 8 dBi. The RDRA is fabricated with Rogers RT Duroid 6010, which offers a permittivity of 10.2.

The concept of MIMO and diversity is extended to the CDRA also as is demonstrated in [9]. The proposed CDRA offers two-port diversity with omnidirectional horizontally and vertically polarized radiation patterns with low cross-coupling. The horizontal and vertical polarization is achieved by exciting the $TE_{01}\delta$ and $TM_{01}\delta$ orthogonal modes, respectively. To improve the impedance bandwidth, cavity perturbation with an air gap in annular ring form is deployed on the CDRA. The proposed antenna provides an impedance bandwidth of 19.1% in the vertically polarized mode and an overlapping bandwidth of 7.4% in the horizontally polarized mode, with the overlapping band ranging from 3.78 to 4.07 GHz. In continuation, [10] explores the concept of MIMO by exciting two orthogonal modes, $HE^{y}_{11\delta}$ and $HE^{x}_{11\delta}$. For this, a suitable microstrip feed network offering orthogonal polarization and decoupling between the ports is designed. The fractional bandwidths for port 1 and port 2 are 17.8% (3.1–3.68-GHz) and 18.4% (3.1–3.7-GHz), respectively, and isolation between the two ports exceeds –25 dB within the required band. Chamfered DRAs aid in generating degenerate orthogonal modes. This concept has been well explored for CP applications. As an extension of this concept, in [11], the researchers excite a chamfered RDRA with an aperture feed and dual ports to achieve MIMO diversity. The antennas are fabricated on Rogers RT Duroid substrate with an Eccostock-made DRA placed over the substrate. The DRAs are excited by an aperture-coupled feed to achieve wide bandwidth and high efficiency. The measured impedance bandwidth of the dual-port MIMO DRA is 27.5% (4.7–6.2 GHz). The maximum gain achieved by the antenna is 7.4 dBi.

In [12], a cross-shaped DRA is made of two different DR slabs of different dimensions but derived from the same material. This DRA configuration is excited by the slot and coaxial port leading to the MIMO application. The rectangular slot generates $TE^x_{\delta 11}$ and $TE^x_{\delta 13}$ orthogonal modes. The coaxial port generates $TE^y_{1\delta 1}$ and $TE^y_{1\delta 3}$ orthogonal modes. Figure 5.3 shows this configuration. The proposed antenna operates at the 3.5-GHz and 5.25-GHz bands. High port isolation is achieved using a hybrid feeding mechanism that excites two orthogonal modes at each frequency band. The measured impedance bandwidth of the proposed antenna covers the entire WiMAX (3.4–3.7-GHz) and WLAN (5.15–5.35-GHz) bands. Very high isolation is achieved between the ports by this method.

In [13], a compact four-port CDRA design is explored for MIMO applications. The proposed CDRA approach is based on using CDRA symmetry with the help of image theory to achieve the best size reduction of the resonators and to maintain the resonance frequency of the original CDRA. The electric/magnetic walls approach is utilized to miniaturize the size by exploring the symmetry and antisymmetry of the resonant mode. First, a CDRA for the MIMO system is designed and tested in terms of return loss and radiation efficiency. Then, two configurations of MIMO antennas (two- and four-port) are examined by using the same substrate size. The two-port MIMO antenna is built from two half-CDRs facing each other. Similarly, four-quarter CDRs are created to form a four-port MIMO antenna system. As a result, a 75% size reduction is achieved (a size of $30 \times 30 \times 7.62$ mm^3) of size. The measured impedance bandwidth for the four-port MIMO antenna is 5.4% (5.4–5.7-GHz), with more than 15-dB isolation levels. Proper levels of ECCs are also achieved (1×10-2-4×10-2), with a CCL of 0.04 bits/S/Hz. This is discussed in [13]. In [14], a similar approach using orthogonal aperture feed lines to an RDRA to achieve MIMO applications as applicable to LTE/4G and prospective 5G is explored. An L-shaped slot is used to achieve orthogonality of the ports to the DRA, thereby generating two-port MIMO applications.

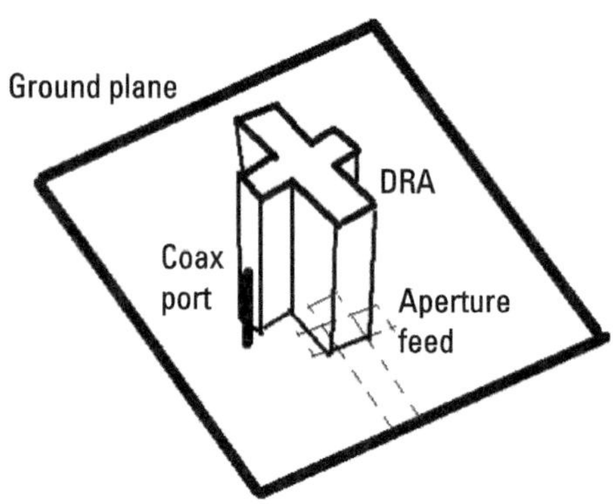

Figure 5.3 Cross-shaped MIMO DRA [12].

5.3 Dual-DRA Elements with Dual- or Multiple-Feeding Techniques for MIMO Applications

Another new technique involves using dual-DRA elements with dual- or multiple-feeding to design various MIMO antennas for various wireless applications. This section explores various such techniques [15–24]. In [15], a hybrid MIMO CDRA is designed for the LTE2500, WLAN, and WiMAX bands. Two CDRAs excited with two symmetric folded microstrip feed lines that are fixed at two different ends of substrate are deployed for achieving triple-band operation. Two inverted L-shaped strips are etched on the ground plane for enhancing the performance of antenna, offering excellent isolation ($S_{12} < -15$ dB) and a 2.7-GHz frequency bandwidth. In addition, to improve the isolation between the two antennas (S12 < –20 dB), metallic strips are created on the ground plane to act as an EM reflector. This proposed MIMO antenna, which has been fabricated and experimentally validated, operates at the 2.24–2.38-GHz, 2.5–3.26-GHz, and 4.88–7.0-GHz frequency bands. Additional folded microstrip lines create a ($\lambda/4$) path delay between the E field lines and generate CP characteristics from the 5.55–5.75-GHz frequency band. Figure 5.4 shows the fabricated prototype of the proposed structure. To fabricate the antenna, two CDRAs of alumina material [Al_2O_3($\varepsilon_r = 9.8$, tanδ 5 0.002)] are fixed on an FR-4 substrate. A folded microstrip line along with modified ground plane is etched on both sides of substrate. This hybrid MIMO antenna is suitable for various applications including LTE2500 (2.5–2.57 GHz for uplink; 2.62–2.69 GHz for downlink), WLAN (5.15–5.35/5.725–5.825 GHz), and WiMAX (5.250–5.850 GHz).

In [16], a simple decoupling method of using metallic vias to improve the isolation of MIMO DRA elements is investigated for millimeter-wave application. A 1 × 2 MIMO DRA array with two different orientations (an H-plane and an E-plane) is designed, and its antenna characteristics are examined. Vertical insertion of vias inside the DRA elements at a proper position creates an interaction with the EM fields that potentially affects the field distributions

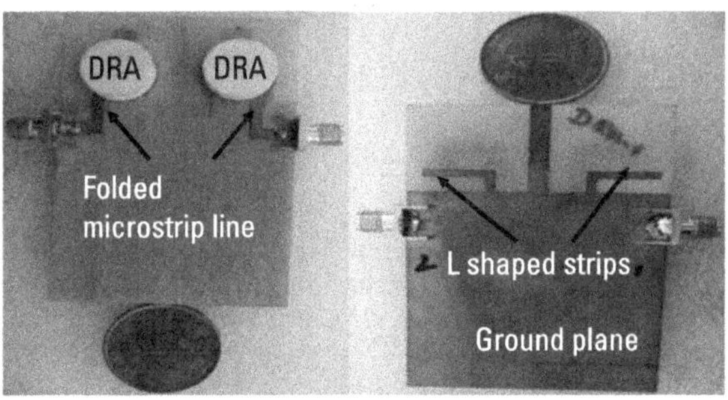

Figure 5.4 Proposed antenna fabricated prototype structure [15].

in the DRA. It also reduces the coupled fields effectively and decreases the mutual coupling between the DRAs. The isolation of a MIMO DRA can be significantly improved by optimized positioning of vias. The isolation of the H-plane-coupled MIMO DRA array improved from ~15.2 to 34.2 dB, and the E-plane array improved from ~13.1 to 43 dB, both at a 26-GHz frequency. This structure offers a compact size with a high isolation level. Also it is applicable for millimeter-wave 5G applications. Figure 5.5 depicts the two different 1 × 2 MIMO DRA arrays (H-plane and E-plane orientation) with their via-loaded structures.

Similarly, in [17], two RDRA arrays placed on a substrate excited by rectangular microstrip-fed slots underneath the dielectric resonator, are investigated [17] for enhancement of the isolation. Each DRA has a top-loaded metal strip that moves the strongest part of the coupling field away from the exciting slot to improve the isolation between the two antenna elements. A 12-dB isolation is achieved over the 27.5–28.35-GHz frequency band. Figure 5.6 shows the geometry of the proposed structure. In [18], a CPW-fed multidielectric permittivity with a stair-shaped dual-RDRA and a metallic diagonal stub is created between the DRAs. For wideband MIMO applications, a new A-shaped dual-element DRA excited by a conformal strip is discussed in [19]. Combining two adjacent modes, namely TM_{101} and TM_{103}, offers a wide bandwidth for various wireless applications. Impedance bandwidths of 59.2% and 60.9% are achieved corresponding to port 1 and port 2, respectively.

Some modified novel techniques are also elaborated for MIMOs with a CP application. In [20], a MIMO antenna is constructed from two identical DRAs that are excited by a cross-slot feeding technique to generate circularly polarized fields. Four metal strips are printed on the lateral sides of each DRA to change the orientation of the E field, which is coupled with the passive (coupled) DRA. CP occurs when the E field in the passive element is orthogonal

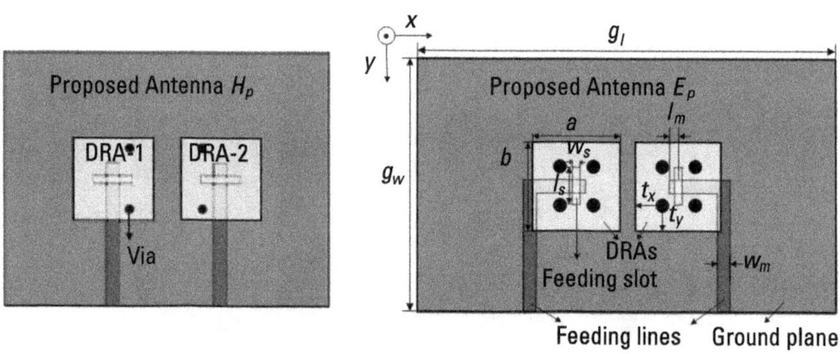

Figure 5.5 Two different 1 × 2 MIMO DRA arrays (an H-plane and an E-plane orientation) with their via-loaded structures [16].

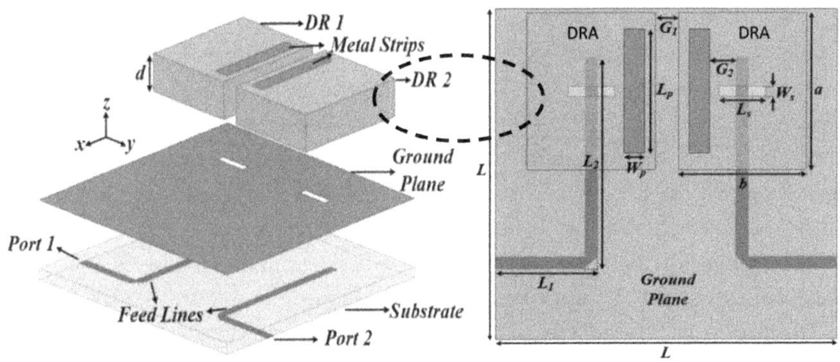

Figure 5.6 Geometry structure of the proposed MIMO DRA [17].

to the active (driven) DRA element. Metal strips play an important role in controlling the mutual coupling between the DRAs. Through polarization orthogonality, the isolation between the two CP DRAs is improved significantly. This proposed antenna configuration, designed for the 2.4-GHz (2.38–2.52-GHz) WLAN band application, achieved a 31-dB maximum isolation. Figure 5.7 illustrates its proposed antenna geometry.

Sequentially, in [21], a dual-band circularly polarized MIMO DRA is investigated for the 3.5- and 5.5-GHz bands. The structure consists of two ring DRAs, shaped and fabricated by RT Duroid dielectric material. CP is obtained by exciting orthogonal HE modes. Placing two probes orthogonally to each other and varying the lengths of the probes, creates a quadrature time-phase

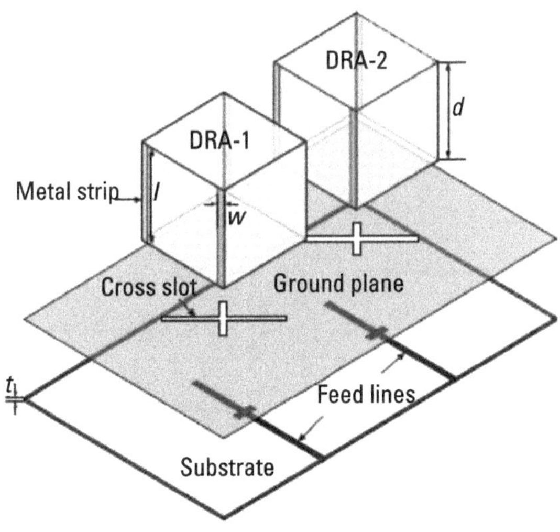

Figure 5.7 Proposed antenna geometry configuration [20].

delay between the modes for achieving CP. Subsequently, in [22], a wideband circularly polarized MIMO DRA is proposed for worldwide interoperability for microwave wireless access (WiMAX) applications. Mutual coupling (MC) between the two DRA elements is significantly reduced by using a new hybrid technique. This hybrid technique employs a parasitic patch at an optimized distance from the conformal metal strip side of the two identical rectangular DRAs, which generates a CP wave with a wide impedance bandwidth. Mutual coupling is diminished significantly when circularly polarized DRAs are placed diagonally at an optimized position. Figure 5.8 shows the geometry of the proposed wideband circularly polarized MIMO antenna, which offers a 20.82% (3.58–4.40-GHz) CP bandwidth and a ~38.51% (3.50–4.95-GHz) impedance bandwidth. Moreover, a –28 dB mutual coupling is achieved over the entire band.

An effective technique for reducing the mutual coupling between DRAs is discussed in [23]. A metasurface is incorporated between the two DRAs, which are placed along the H-plane for improvement of isolation. The metasurface consists of an array of SRR cells, which are integrated along the E-plane. The SRR structure is designed for offering a bandstop property within the antenna operating frequency range, which substantially reduces the mutual coupling between the adjacent dielectric radiators. A 28-dB reduction in the mutual coupling level is achieved by loading the DRA with 1×7 array of SRR unit cells without degrading the antenna performance. Figure 5.9 illustrates the structure of the 1×2 DRA with 1×7 MTM-based SRR in the H-plane. Similarly, in [24], a MIMO DRA's isolation is improved by using a dielectric superstrate. The superstrate affects the coupled field distributions inside the adjacent DRAs, thereby minimizing the coupling between two closely positioned DRAs. The proposed techniques offer a unique approach for isolation enhancement of the MIMO DRAs.

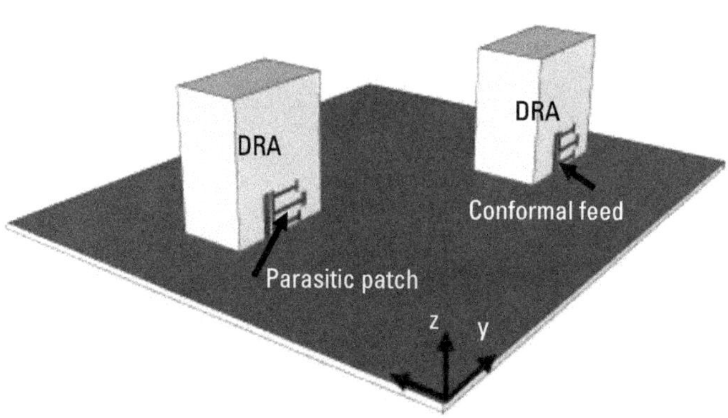

Figure 5.8 Geometry of proposed wideband circularly polarized MIMO antenna [22].

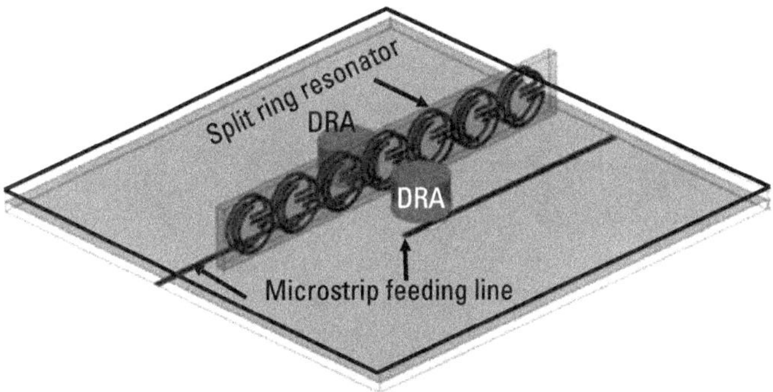

Figure 5.9 Structure of 1 × 2 DRA with 1 × 7 MTM-based SRR in the H-plane [23].

5.4 Multiple DRA Elements with Multiple Feeds for MIMO Applications

Leading-edge wireless technologies need MIMO antenna systems, which offer higher data rates, larger system capacities, and wider bandwidths than single-antenna systems. This section explores multiple DRA elements with multiple feeding techniques for various wireless technologies [25–28]. In [25], a compact six-port DRA array is designed for polarization, angle diversities, and enhancement of the channel capacity; the 6 × 6 antenna system operates at a 2.65-GHz resonant frequency for an indoor office service system. This antenna offers a 1.5-bits/s/Hz channel capacity for wireless applications. In [26], a compact back-to-back DRA-based four-port MIMO antenna system with bi-directional diversity is investigated for WLAN applications. The back-to-back arrangement of CDRAs is one of the unique characteristics of the proposed antenna configuration. The CDRAs are placed on both sides of an FR4 substrate with a common ground plane. Each element of the CDRA is excited by two ports to demonstrate a four-port MIMO antenna system. The upper side of CDRA is excited by CPW-fed conformal strip lines at port 1 and port 2. Similarly, the lower side of the CDRA is excited by microstrip line–fed conformal strip lines at ports 3 and 4. Isolation between all ports is enhanced by the generation of two orthogonal modes in the CDRAs and the use of two orthogonal feed excitations. Figure 5.10(a) shows the configuration of the proposed antenna. Two alumina ceramic-based CDRAs ($\varepsilon_{dra} = 9.8$) DRA-1 and DRA-2 are placed on the opposite sides of an FR4 substrate ($\varepsilon_{sub} = 4.4$) sharing a common ground plane. The isolation is improved between ports 1 and 2, when ($HE^{x}_{11\delta}$ and $HE^{y}_{11\delta}$) orthogonal modes are excited corresponding to DRA-1. Figure 5.10(b) depicts the bidirectional radiation diversity between various ports. A single CDRA element with dual-port excitation offers two orthogonal radiation patterns. This antenna attains good isolation, an above 5-dBi gain with 0.25 below ECC at all corresponding ports.

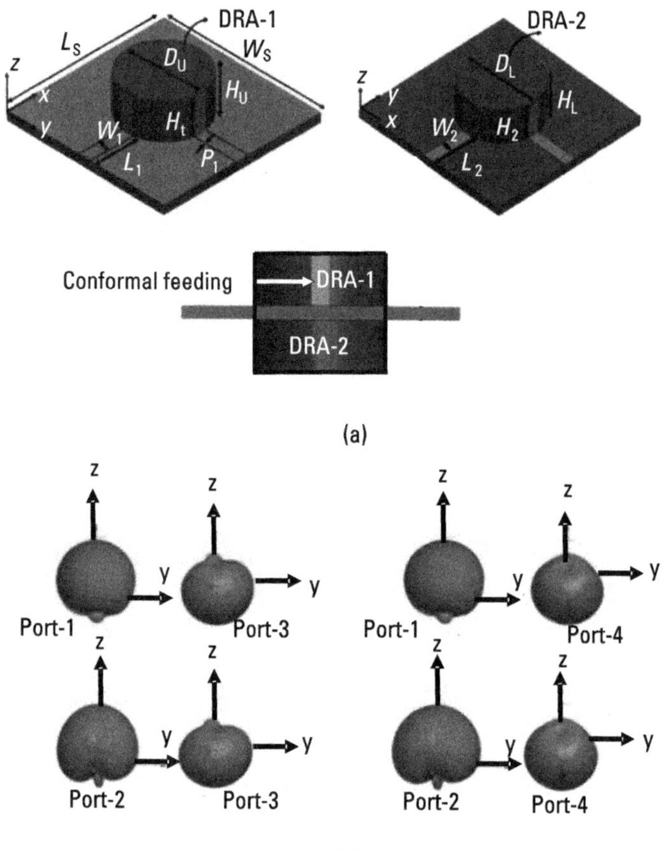

Figure 5.10 Proposed antenna: (a) configuration of the proposed antenna and (b) bidirectional radiation diversity between various ports [26].

A three-dimensional dual-band DRA is investigated for various wireless applications in [27]. This MIMO antenna system consists of five rectangular dielectric resonator radiating elements. The proposed MIMO antenna or diversity antenna system used $TE^{y}_{1\delta 1}$ and $HEM_{2\delta 1}$ modes for the transmission or reception of data from various directions. The $TE^{y}_{1\delta 1}$ mode covers the 3.3–3.8-GHz frequency range for the Wi-MAX, sub-6-GHz, and LTE (3.41–3.5-GHz uplink/3.51–3.6-GHz downlink) 5G application bands. Similarly, the $HEM_{2\delta 1}$ mode covers wi-fi (5.725–5.875-GHz) and ITS 5.9-GHz (5.875–5.925-GHz) wireless bands. RDRA elements are excited by coaxial feeds, which are arranged in a cubical order to cover maximum directions for radiation. This antenna geometry offers a more than 7-dBi gain in the lower sideband and a 7.2-dBi gain in the upper side frequency band. Figure 5.11 depicts this three-dimensional MIMO antenna geometry.

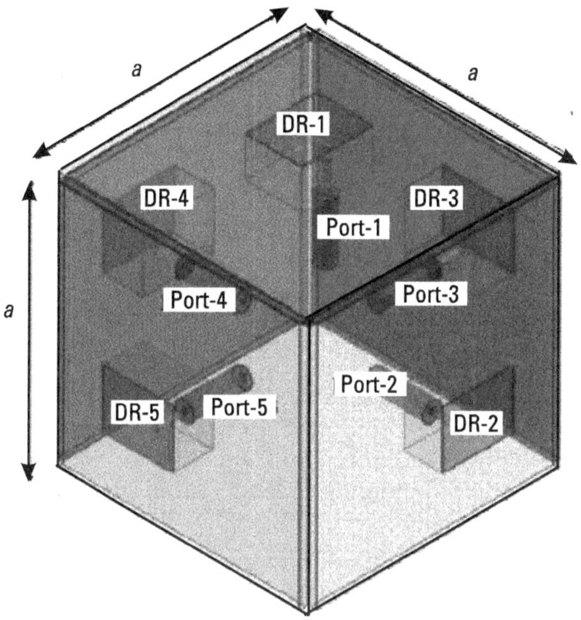

Figure 5.11 Three-dimensional MIMO antenna geometry [27].

Similarly, an integrated four-port MIMO system with CDRA is designed for a cognitive radio (CR) application in [28]. The proposed antenna performs an ultrawideband (UWB) and a narrowband (NB) dual-state operation, which is achieved through a modified feeding technique. For the UWB state, the antenna consists of two-port circular dielectric radiators and is excited by a microstrip feed line, while for NB state, two CDRAs are integrated with UWB radiators. In [29], a four-element, eight-port CDRA with multidirectional pattern diversity is investigated for wireless access point applications. A dielectric resonator is excited by two different feeding techniques to create two orthogonal radiation patterns. To design an eight-port MIMO antenna system, a primary dielectric radiator is combined with similar types of radiators in a cross-configured manner. Antenna elements are arranged in such an order, that each radiator of the proposed MIMO system can radiate in a different direction, thus minimizing the field correlation among them. The isolation between the antenna ports is enhanced by generating orthogonal modes in each CDRA, thereby exciting each CDRA to radiate in a different direction. Figure 5.12 shows this four-element, eight-port CDRA configuration. The antenna offers more than 20 dB of isolation between the ports and multidirectional pattern diversity. As well, less than 0.25 ECC with a 5-dBi average gain is attained. The antenna covers the 5.6–s5.9-GHz bands for WLAN standard wireless application.

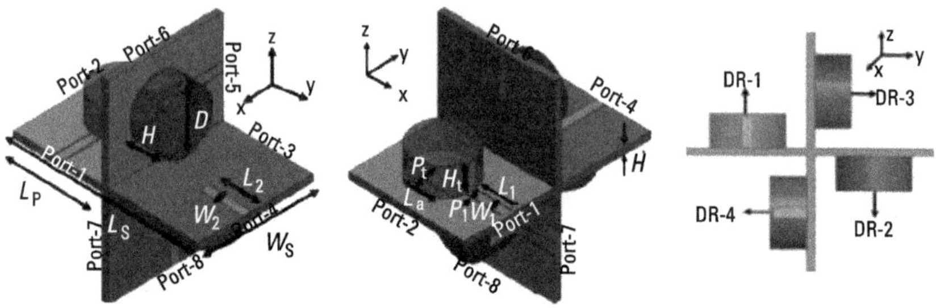

Figure 5.12 Four-element, eight-port CDRA configuration [29].

Some other new MIMO antenna design techniques are discussed in [30–32]. In [30], a triple-port, two-mode-based two-element CDRA for MIMO applications is investigated. In [31], an eight-element, 16-port dielectric resonator–based, double-sided multiport antenna is investigated to achieve unique dual-directional pattern diversity. Similarly, an FSS-based four-port MIMO CDRA system contributes to multidirectional pattern diversity characteristics in [32]. The back-to-back arrangements of the CDRA and the incorporation of FSS improves the isolation between the ports.

5.5 Conclusion

This chapter discussed MIMO antenna characteristics in terms of the reflection coefficient, transmission coefficient, ECC, CCL, and MEG parameters. Dual- or multiple-feeding techniques with single, dual, and multiple DRA elements were discussed for MIMO antenna system design. The chapter also detailed some new techniques, orthogonal feeding techniques and orthogonal mode generation, for improving isolation and pattern diversity.

References

[1] Malviya, L., R. K. Panigrahi, and M. V. Kartikeyan, *MIMO Antennas for Wireless Communication,* Boca Raton, FL: CRC Press, 2021.

[2] Sharawi, M. S., *Printed MIMO Antenna Engineering,* Norwood, MA: Artech House, 2014.

[3] Aouadi, B., and J. B. Tahar, "Investigation of Capacity Loss and Mean Effective Gain in Multiple Antenna Systems," *IEEE 10th International Conference on Wireless and Mobile Computing, Networking and Communications (WiMob),* 2014, pp. 1–6.

[4] Yan, J., and J. T. Bernhard, "Design of a MIMO Dielectric Resonator Antenna for LTE Femtocell Base Stations," *IEEE Transactions on Antennas and Propagation,* Vol. 60, No. 2, Feb. 2012, pp. 438–444.

[5] Yan, Y., and J. T. Bernhard, "Implementation of a Frequency-Agile MIMO Dielectric Resonator Antenna," *IEEE Transactions on Antennas and Propagation,* Vol. 61, No. 7, July 2013, pp. 3434–3441.

[6] Roslan, S. F., et al., "An MIMO Rectangular Dielectric Resonator Antenna for 4G Applications," *IEEE Antennas and Wireless Propagation Letters,* Vol. 13, 2014, pp. 321–324.

[7] Karimian, R., et al., "Low-Mutual-Coupling 60-GHz MIMO Antenna System with Frequency Selective Surface Wall," *IEEE Antennas and Wireless Propagation Letters,* Vol. 16, 2017, pp. 373–376.

[8] Abdalrazik, A., A. S. A. El-Hameed, and A. B. Abdel-Rahman, "A Three-Port MIMO Dielectric Resonator Antenna Using Decoupled Modes," *IEEE Antennas and Wireless Propagation Letters,* Vol. 16, 2017, pp. 3104–3107.

[9] Zou, L., D. Abbott, and C. Fumeaux, "Omnidirectional Cylindrical Dielectric Resonator Antenna with Dual Polarization," *IEEE Antennas and Wireless Propagation Letters,* Vol. 11, 2012, pp. 515–518

[10] Das, G., A. Sharma, and R. K. Gangwar, "Dual Port Aperture Coupled MIMO Cylindrical Dielectric Resonator Antenna with High Isolation for WiMAX Application," *International Journal of RF and Microwave Computer-Aided Engineering,* Vol. 27, No. 7, 2017.

[11] Singhwal, S. S., et al., "Dual-Port MIMO Dielectric Resonator Antenna for WLAN Applications," *International Journal of RF and Microwave Computer-Aided Engineering,* 2019.

[12] Khan, A. A., Khan, R., Aqeel, S., Nasir, J., Saleem, J., & Owais., "Design of a dual-band MIMO dielectric resonator antenna with high port isolation for WiMAX and WLAN applications," *International Journal of RF and Microwave Computer-Aided Engineering,* Vol. 27, No. 2, 2016.

[13] Belazzoug, M., et al., "Ultra-Compact 4-Port DR Antenna for Multiinput Multioutput Standards," *International Journal of RF and Microwave Computer-Aided Engineering,* 2020.

[14] Anuar, S. U., et al., "Triple Band MIMO Dielectric Resonator Antenna for LTE Applications," *AEU—International Journal of Electronics and Communications,* Vol. 118, 2020, pp. 153172.

[15] Sharma, A., G. Das, and R. K. Gangwar, "Dual Polarized Triple Band Hybrid MIMO Cylindrical Dielectric Resonator Antenna for LTE2500/WLAN/WiMAX Applications," *International Journal of RF and Microwave Computer-Aided Engineering,* Vol. 26, No. 9, 2016, pp. 763–772.

[16] Pan, Y. M., et al., "A Simple Decoupling Method for 5G Millimeter-Wave MIMO Dielectric Resonator Antennas," *IEEE Transactions on Antennas and Propagation,* Vol. 67, No. 4, April 2019, pp. 2224–2234.

[17] Zhang, Y., et al., "A MIMO Dielectric Resonator Antenna with Improved Isolation for 5G mm-Wave Applications," *IEEE Antennas and Wireless Propagation Letters,* Vol. 18, No. 4, April 2019, pp. 747–751.

[18] Sani, M. M., R. Chowdhury, and R. K. Chaudhary, "An Ultra-Wideband Rectangular Dielectric Resonator Antenna with MIMO Configuration," *IEEE Access,* Vol. 8, 2020, pp. 139658–139669.

[19] Sharma, A., et al., "A-Shaped Wideband Dielectric Resonator Antenna for Wireless Communication Systems and Its MIMO Implementation," *International Journal of RF and Microwave Computer-Aided Engineering,* 2018.

[20] Hu, Y., Y. M. Pan, and M. Di Yang, "Circularly Polarized MIMO Dielectric Resonator Antenna with Reduced Mutual Coupling," *IEEE Transactions on Antennas and Propagation.*

[21] Singhwal, S. S., et al., "Dual-Band Circularly Polarized MIMO DRA for Sub-6 GHz Applications," *International Journal of RF and Microwave Computer-Aided Engineering,* 2020.

[22] Iqbal, J., et al., "Mutual Coupling Reduction Using Hybrid Technique in Wideband Circularly Polarized MIMO Antenna for WiMAX Applications," *IEEE Access,* Vol. 7, 2019, pp. 40951–40958.

[23] Dadgarpour, A., et al., "Mutual Coupling Reduction in Dielectric Resonator Antennas Using Metasurface Shield for 60-GHz MIMO Systems," *IEEE Antennas and Wireless Propagation Letters,* Vol. 16, 2017, pp. 477–480.

[24] Hu, Y., Y. M. Pan, and M. Di Yang, "Circularly Polarized MIMO Dielectric Resonator Antenna With Reduced Mutual Coupling," *IEEE Transactions on Antennas and Propagation.*

[25] Tian, R., et al., "A Compact Six-Port Dielectric Resonator Antenna Array: MIMO Channel Measurements and Performance Analysis," *IEEE Transactions on Antennas and Propagation,* Vol. 58, No. 4, April 2010, pp. 1369–1379.

[26] Das, G., et al., "Compact Back DRA-Based Four-Port MIMO Antenna System with Bidirectional Diversity," *Electronics Letters,* Vol. 54, No. 14, 2018, pp. 884–886.

[27] Mishra, S., et al., "Three-Dimensional Dual-Band Dielectric Resonator Antenna for Wireless Communication," in *IEEE Access,* Vol. 8, 2020, pp. 71593–71604.

[28] Pahadsingh, S., and S. Sahu, "Four Port MIMO Integrated Antenna System with DRA for Cognitive Radio Platforms," *AEU—International Journal of Electronics and Communications,* Vol. 92, 2018, pp. 98–110.

[29] Gangwar, R., et al., "Dielectric Resonator Based 4-Element 8-Port MIMO Antenna with Multidirectional Pattern Diversity," *IET Microwaves, Antennas & Propagation,* 2018.

[30] Das, G., et al., "Triple-Port, Two-Mode Based Two Element Cylindrical Dielectric Resonator Antenna for MIMO Applications," *Microwave and Optical Technology Letters,* Vol. 60, No. 6, 2018, pp. 1566–1573.

[31] Das, G., N. K. Sahu, and R. K. Gangwar, "Dielectric Resonator Based Multiport Antenna System with Multidiversity and Built-In Decoupling Mechanism," *AEU—International Journal of Electronics and Communications,* Volume 119, 2020, pp. 153193.

[32] Das, G., et al., "FSS-Based Spatially Decoupled Back-to-Back Four-Port MIMO DRA with Multidirectional Pattern Diversity," *IEEE Antennas and Wireless Propagation Letters,* Vol. 18, No. 8, Aug. 2019, pp. 1552–1556.

6

Fabrication, Experimental Setup, Measurement, and Practical Considerations

This chapter details the various practical considerations one needs to make before performing the DRA fabrication, testing, and measurement processes. In addition, since researchers are often confused when choosing the dielectric material for their applications, the chapter describes Ashby's material selection method for selecting the appropriate dielectric material based on specified requirements. The chapter also discusses the fabrication mechanism and measurement setup needed for DRA measurements.

6.1 Dielectric Material and Its Selection

The choice of a correct dielectric material depends on a number of different features. Since DRAs are modeled as cavity resonators, their resonant frequency is determined by the physical dimensions and permittivity of the material chosen and its ambience where the resonator is placed. For example, shielded DRAs will have a different resonant frequency than the open-ended DRAs.

For the choice of a material for the DRA, the key properties include relative permittivity (ε_r), quality factor (Q factor), and near-zero temperature coefficient of the resonant frequency (τ_f). As the dielectric materials used generally for DRA applications have a high permittivity, it is expected that the Q factor of such materials shall also be high. According to the literature, researchers have used several dielectric materials with a variation in permittivity from 4 to 27.

The leading manufacturers of DRA materials include Rogers Corporation, Eccostock, and Taconic [1–3].

Glass-based DRAs have also been explored for various applications and can be suitably excited using slot feed, aperture feed, or LEDs to achieve tunability [4–6]. In fact, it is interesting to note that optical fibers are cylindrical dielectric waveguides based on silica or glass material only, operating at the terahertz and higher frequency ranges. In addition, ceramic-based oscillators and filters are well-known.

However, at microwave frequencies, mapping of the DRAs based on applications has been well described in [7]. This is based on Ashby's method, though other methods like multiattribute utility analysis, graph theory, and the multiple-attribute decision-making approach may also be used [8–11].

There are two important parameters of concern based on Ashby's method: the material indices and the performance indices. This is detailed by several researchers in [11–13]. The optimization and selection of the most appropriate dielectric material is based on the materials' properties, which are known as material attributes (material indices and performance indices). These material indices define the constraints and affect the performance of materials in terms of performance indices based on the defined objectives.

The various steps involved in the Ashby's methodology are described as follows [11–13]:

1. *Translation:* Listing a class of available materials with their attributes as a function of constraints and objectives.
2. *Screening:* Eliminating the materials that do not fulfill the objective, as per the available material indices and their respective performances.
3. *Ranking:* Ranking the screened material to satisfy the performance indices.
4. *Supporting information:* Exploring pedigrees of the top-ranked candidates or materials.

The top ranked materials show the best performance indices and are selected by Ashby's approach.

In the Ashby approach to material selection [11], a function is sought to describe the performance (p) of the element under consideration.

In general, this function has the form $p = \{F, G, M\}$ where F, G, and M express the functional requirements, geometric parameters, and material indices, respectively. These steps are well elaborated in [7].

The choice of material indices is based on the following dielectric material parameters:

- Relative permittivity (ε_r);
- Quality factor (Q factor);
- Near-zero temperature coefficient of the resonant frequency (τ_f).
- Any material manufacturer always provides this information on their data sheets. In fact, the material indices are a first step to designing any geometry or shape of DRA; subsequently, the performance indices are either chosen or specified. The various performance indices for the material selection are listed as follows.
- Resonant frequency (f_0): The resonant frequency is a function of the dimensions, shape, and effective permittivity of the material. The calculations of the resonant frequency are well explained in previous chapters and in [14]. The resonant frequency also depends on the kind of mode that can be excited for a clearly defined geometry. As an example, we consider the parameters used in the calculation of the resonant frequency for the RDRA, the DWM, listed as follows.
- Size of the antenna: The electrical size or electrical length of the antennas based on the wavelength or λ. Since the physical sizes can be in centimeters or millimeters, it is inappropriate to compare the size using a physical scale. Thus, electrical lengths give a fair idea of the DRA compatibility's with the other circuits to be integrated.
- Impedance bandwidth (BW): Calculated from the scattering parameters (S-parameters). For an antenna it is defined as the frequency range for $|S_{11}|<=-10$ dB. In terms of the VSWR, the bandwidth spans for a frequency range for VSWR< = 2. The relation between the bandwidth and Q factor is given as follows [14]

$$BW = \frac{VSWR - 1}{Q\ factor \times \sqrt{VSWR}} \tag{6.1}$$

where the Q factor is given as

$$Q\ factor \sim \frac{1}{\tan\delta} \propto \varepsilon_r^{\frac{3}{2}} \tag{6.2}$$

- Radiation efficiency (η): The capability of an antenna to convert the input radio wave power accepted at its terminals (P_{input}) into radiated power (P_{radiated}), shown as follows

$$\eta = \frac{P_{radiated}}{P_{input}} \tag{6.3}$$

It is also defined as the ratio of the radiation resistance to its total resistance (i.e., the radiation resistance and the loss resistance taken together).

For a DRA, since there is no separate metallic intrusion, the radiation efficiency is expected to be high. Also, unlike the planar microstrip antenna counterparts, where the edges of the radiator radiate, in a DRA the whole bulk of the material surface contributes to radiation. This is also evident from the fact that the EM modes of a DRA are a function of volume and so radiation efficiency of the DRA is expected to be high.

Based on the information gathered about material and performance indices, the researchers in [7], perform the optimization and plot three graphs (shown in [7]), described as follows:

- Loss tangent (tanδ) against the relative permittivity (ε_r): Used to study those materials that maximize the radiation efficiency (η) of the antenna. It is considered that a material with moderate values of permittivity and much less loss tangent value is suitable to minimize the size of the antenna and maximize the bandwidth. This plot helps to identify a lower value of tan(δ) as compared to other materials with similar permittivity values; however, some materials possess a high value of near-zero temperature coefficient of the resonant frequency (τ_f).
- Loss tangent (tanδ) against the temperature coefficient of the resonant frequency (τ_f) measured in ppm/°C: This plot helps to identify those materials that maximize the radiation efficiency (η) of the antenna.
- Temperature coefficient of the resonant frequency (τ_f) against the permittivity (ε_r): It is found that a material with moderate values of permittivity and a near-zero temperature coefficient of the resonant frequency (τ_f) is suitable to minimize the size of the antenna and maximize the radiation efficiency (η). This also aids in improving the bandwidth of operation.
- Based on the three plots, the material selection of the desired permittivity is made for the suitable application. For example, in order to fulfill the desired criteria of dielectric resonator material for the DRA, the results show that Roger TMM10 ($\varepsilon_r = 9.8$) is the best possible material out of all the materials taken into consideration for various wireless communication applications at the microwave-frequency range [7]. FR 4-based DRAs have also been explored, because FR 4 is relatively cheaper than other materials; however, since FR 4 is lossy, it constrains the design of

high-efficiency DRAs. FR 4 laminates are mostly used for providing slot or aperture or microstrip feed to the DRAs.

6.2 Limitations in Practical DRA Considerations

One major drawback of DRAs is fabrication issues, and therefore the machining from various mechanical laboratories is now included to minimize the issues of fabrication tolerances during experiments and measurements. However, it is interesting to categorize the frequency limitations of the DRA as follows:

1. Lower-frequency bound: This is mostly attributed to the size and the weight considerations of the DRA. As an example, a dielectric material of permittivity $\varepsilon_r = 9.2$, a dissipation factor $\tan\delta = 0.0022$, and a density of 2.8 gm/cm^3 is considered. For this material, an HDRA with radius $r = 2.54$ cm (1 inch) when given an offset probe feed at $x = 1.74$ cm and a length of probe at 1.52 cm will resonate at 1.816 GHz offering a 10.5% impedance bandwidth [15].

   When compared to a microstrip square patch antenna with dimensions $L_p = W_p = 75$ mm symmetrically located above the ground plane; and a thickness of substrate of 6mm, dielectric constant of 2.62 and loss tangent 0.001 resonates at 2 GHz [16]. While microstrip patch is a two-dimensional geometry, the HDRA is a three-dimensional geometry, and so it is clearly understood that DRAs suffer from size and weight issues in the lower frequency bounds.

2. High-frequency bound: The high-frequency applications are also bounded by size and the feeding techniques used to excite the DRA. However, at very high frequencies, the DRAs offer a major advantage over the patch antennas. The ohmic losses at such frequencies become nonlinear in nature, and since DRAs do not have any metallic inclusion, at such frequencies, DRAs are less lossy.

   Nevertheless, the feeding techniques become crucial. For example, a probe feed at such frequencies is a major challenge. It is difficult to achieve such a small diameter and low lossy ohmic wires for a probe. Thus, a probe feed is preferred for low frequencies due to its compact size. Microstrip feed lines and aperture coupling present a limitation since their size and dimensions are limited by fabrication technologies.

It is also interesting to note that the size of the DRA is proportional to $\frac{\lambda_0}{\sqrt{\varepsilon_r}}$, where λ_0 is the operating wavelength and the relative permittivity is given by ε_r. Through this proportionality, the size of the DRA can be easily scaled over any given frequency range. As an example, if operating at a lower frequency range, sometimes the size of the DRAs tend to become very large. Then by choosing a high-permittivity material in the scaling factor, the size of the DRA can be significantly reduced. The same holds valid for linear proportionality with the operating wavelength as well. So, a microwave-operating DRA can also be escalated to operate at the millimeter-wave or terahertz range, in a situation that is better known in the research domain as THz dot antennas [17]. In addition, at such frequencies, the radiation efficiency of DRAs is high due to the absence of any surface wave propagation, unlike in its planar antenna counterparts.

At very high frequencies, low-temperature co-fired ceramics (LTCCs) may be a solution, though they may offer a limited range of effective permittivity. As well, while they tend to be lossy, appropriate sizes of DRAs for high frequencies can be manufactured. Other possible feed techniques at very high frequencies include direct microstrip line, CPWs, dielectric image waveguides, and proximity-coupled waveguides.

Another significant limitation of DRAs is the fact that most of dielectric slab and laminate manufacturers export their material in fixed slab sizes, which are then subjected to rigorous machining to obtain the desired geometry and dimensions. Thus, frequently, these slabs are glued at the edges and then joined under some high pressure such as a rig to prevent ruptures in the material and the creation of air gaps between the layers of the DRA.

The presence of an air gap can seriously alter the results of the DRA as it tends to decrease the effective permittivity of the material. Similarly, the permittivity of glue applied over the complete surface of the slab for stacking, is also likely to alter experimental results.

Therefore, it is suggested that glue be applied at the edges and that the stacked slabs or laminates be subjected to high pressure for some time to avoid any of these effects.

6.3 Fabrication Methods and Machines

Various researchers have explored numerous methods for realizing DRAs. The fabrication techniques mostly involve machining, unlike PCB fabrication techniques, which need sophisticated etching techniques. Some of the machines used for DRA fabrication are described as follows.

1. Water jets or abrasive jet machines: The abrasive jet machine uses high-pressure gas, air, or water to carry a high-speed and high-pressure stream of abrasive particles or abrasive sand (silica carbide) through a small nozzle. The size of the nozzle decides the smallest notch that

can be created on the dielectric material surface. This is based on a subtractive manufacturing technique, and it erodes away the material from the work piece.

The work piece and the nozzle of the abrasive jet are in perpendicular position to each other. This standard setup gives precise machining for a dielectric job for machining purposes. Usually, a nozzle dimension of 1 mm or even lower is available. Figure 6.1(a) shows the water jet/abrasive jet machine available at PDPM IIITDM Jabalpur. Material is removed by fine abrasive particles of about 0.001 inch or 0.025 mm in diameter. The pressure of the gas (usually inert gases in nature) ranges from 25 to 130 psig, 170–900 kPa, or four bars, and the speeds can reach as high as 300 m/s or 1,000 km/h. Figure 6.1(b) shows various fabricated structures from the abrasive jet machine. The DRAs shown in clockwise order in Figure 6.1(b) are taken from [18–22].

2. Computer numerical control (CNC) milling machine: This machine, based on the subtractive manufacturing technique, is moved and monitored by computer control numerically and so minimizes human error. The CNC's computer software transforms the numerical codes into Cartesian coordinates, providing a higher degree of freedom for the machine to operate just like a robot.

   A CNC milling machine uses a rotating cylinder cutter to move along multiple axes and to create slots, holes, and various other details

(a)

(b)

Figure 6.1 Abrasive water jet machine with high pressure cutting: (a) water jet machine and (b) different DRA structures fabricated in [18–22] in clockwise order.

in the material under inspection. CNC machines can be three, five, or seven axes, depending on the precision needed for the job order. The machined materials can be plastics, ceramics, composite materials, or metals. The vertical orientation of CNC milling usually has long and thin cutting tools, whereas the horizontal orientation has shorter and thicker cutting tools. Figure 6.2 shows a three-axis CNC milling machine. The DRAs fabricated or machined using CNC milling are discussed in [23, 24].

3. Lathe machine: A lathe machine is conventionally defined as a machine tool that removes the undesired material from a rotating work piece in the form of chips with the help of a tool that is traversed across the work piece and that can also be fed deep into the work piece. This is considered one of the oldest possible machines capable of forming shaped objects for bodies of revolution.

   Modern lathe machines can also be controlled through instructions passed by a computer using an appropriate interfacing software. These

Figure 6.2 A three-axis CNC milling machine.

are modern computer-controlled lathe machines for achieving high definition and precision of the job force under consideration.

Figure 6.3(a) shows a conventional lathe machine operated mechanically, and Figure 6.3(b) shows a computer-controlled lathe machine; Figure 6.3(c) depicts the various DRAs fabricated on the lathe machines. The DRAs, referenced from [25–28], are all fabricated from Rogers Corporation ceramic materials and FR 4 material.

4. Laser cutting machine: Laser cutting is a popular technology that uses a high-power laser beam to cut or slice different materials.

   A high-powered laser beam along with a CNC machine are used to direct the material or the laser beam generated. A commercially available laser for shearing materials deploys a motion-control system to follow a CNC or G-code of the pattern to be cut or created on the material.

   The focused laser beam is then directed at the material, which either melts, burns, or vaporizes, or is blown away by a jet of gas, thus leaving an edge with a high-quality surface finish. Usually available in institutes, a laser cutting machine looks like the one shown in Figure 6.4. Thin laminates can be individually cut using the laser cutting machine and then later be combined or stacked by applying glue at the edges under high pressure. See [29] for a detailed discussion of the fabrication of various wearable antennas using a laser cutting machine.

5. Three-dimensional printers (3-D printers): In this additive manufacturing technique, the construction of a three-dimensional object from a CAD model or digital 3-D model is rendered. This is a type of rapid

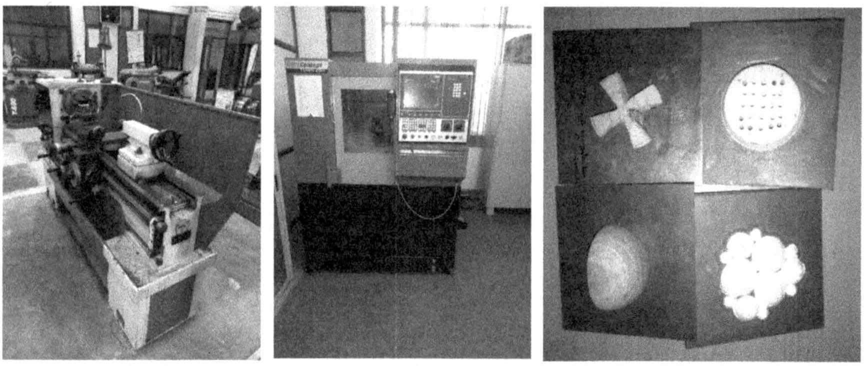

(a) (b) (c)

Figure 6.3 (a) Manually operated lathe machine; (b) a computer-controlled or -digitized lathe machine; and (c) different DRAs fabricated from the lathe machine as referenced in [25–28], in clockwise order.

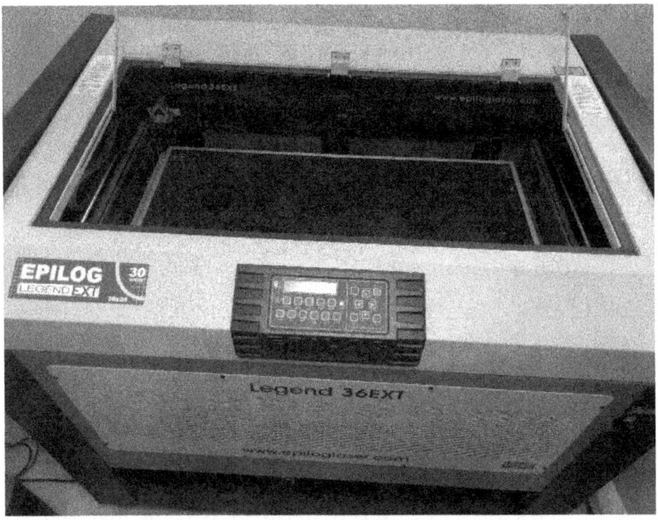

Figure 6.4 A laser cutting machine for cutting thin substrates.

prototyping. In this process, the materials are deposited, joined, or solidified under computer control to render and create the 3-D prototype. Different materials can also be added and can be created layer by layer. Fused-deposition modeling (FDM), which uses a continuous filament of a thermoplastic material, is the most common 3-D printing process in use. Three-dimensional printing has recently been used to achieve various complex shapes for prototyping in DRAs [30–34]. The materials used for prototyping include ceramics and metallic inclusions. Figure 6.5 shows a 3-D printer.

Among all the fabrication or machining methods discussed, fabrication tolerance remains a major constraint. For example, even when the same materials are stacked to obtain the required thickness or height of the material, some air gaps always persist, which changes the effective permittivity of the material and, thus, a shift observed in the resonant frequencies. Also, since the effective permittivity is lowered, the impedance bandwidth of the proposed DRA is also altered. Moreover, if too much adhesive is applied on the DRA surface to stack the materials, it may alter the effective permittivity of the material, skewing experimental results.

Fortunately, with modern simulation tools, including solvers based on FIT, full-wave FEM solvers, and FDTD, such discrepancies can be investigated; subsequently, the simulation and experimental results can be matched to achieve better agreement.

Figure 6.5 A 3-D printer as a tool for additive manufacturing available at the mechanical workshop of IIITDM Jabalpur.

Also note that some small tolerances are inherently present due to the limitation or age of the machines deployed. For example, older water jet machines or abrasive jet machines suffer from increases in the nozzle dimensions. Thus, even though the nozzle of the abrasive jet may be 1 mm, an old nozzle may cut the material with a 1.5-mm dimension, which may affect any slot size that needs to be grooved in the material.

In addition to fabrication tolerances or machining limitations, cable losses, connector losses, connector joints, or splicing may also limits the accuracy of measurements. These are mostly attributed to the measurement setup, a reminder of the importance of taking the utmost care during experiments.

6.4 Various Measurements of the DRA

The measurement techniques deployed for DRA measurements are the same for any other antenna under test (AUT). A typical measurement setup includes a vector network analyzer (VNA) connecting ultra-low loss RF cables, an anechoic chamber or anechoic box setup, and an antenna positioner with a turn table; see Figure 6.6(a–c).

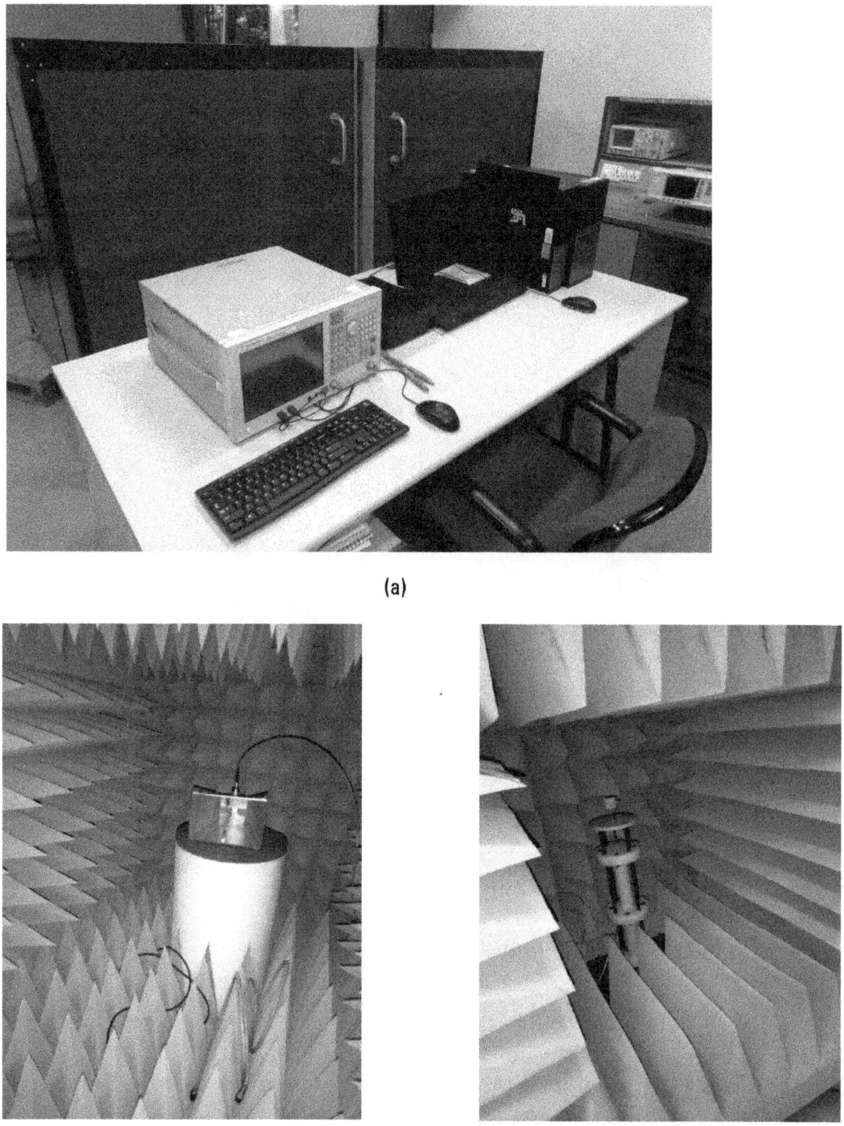

(a)

(b) (c)

Figure 6.6 The testing and measurement setup: (a) setup with VNA, comprising a computer to plot the radiation pattern and anechoic boxes up to a 20-GHz measurement setup. (b) The broadband horn antenna as the transmitter antenna operational up to 16 GHz placed inside the anechoic box. (c) The antenna under test placed on the turntable inside the anechoic box.

Figure 6.6(a) shows the VNA setup, a computer to plot the radiation pattern, RF cables to be connected to the VNA, and anechoic boxes up to a 20-GHz measurement. Figure 6.6(b) shows the broadband horn antenna as the transmitter antenna operational up to 16 GHz placed inside the anechoic box. Figure 6.6(c) shows the antenna under test placed on the turn table inside the anechoic box. The antenna positioner is connected to the computer and the VNA through the anechoic box, which, with customized software, plots the radiation pattern of the AUT. Sections 6.4.1–6.4.4 discuss the key elements of DRA measurement.

6.4.1 Measurement of the S-Parameter

The S-parameter gives the reflection coefficient of the AUT. Through the S-parameter one can easily predict the impedance bandwidth of the AUT. Since the antenna is a one-port device, through VNA only the $|S_{11}|$ or the reflection coefficient is sufficient. The relation RL and the reflection coefficient are given as

$$\mathrm{RL} = -20\log\left(\left|\Gamma_{11}\right|\right) \tag{6.4}$$

where RL is a positive quantity and $|\Gamma_{11}|$ is the magnitude of the reflection coefficient. The reflection generated in the cables connected to the VNA generate the information of the reflection coefficient, which is swept against the frequency on the screen of the VNA.

It is advisable that before the measurement of the reflection coefficient is done, the VNA is essentially calibrated. The VNA calibration is done to remove any system errors from the hardware of the instrument and to take into account the presence of any accessories that may have been added to enable specific measurements to be performed at the required frequency band for the measurements.

The most popular method of VNA calibration is using a Cal-kit, which consists of various loads: a short, open load (usually a precalibrated 50-ohm termination) and a thru. By connecting the loads one by one on the RF calibration, embedding or de-embedding of the RF cable to avoid phase errors in the calibration can also be achieved. The impedance bandwidth of the AUT is measured by the $|S_{11}| \leq -10$ dB line. VNA manufacturers include Keysight Technologies, Rohde & Schwarz, and Anritsu [36–38]. The measurement of the reflection coefficient of the AUT from VNA also reveals the Smith chart plot and the VSWR. In general, for good impedance matching, the reflection coefficient encircles the center of the Smith chart very closely. Also analogous to the impedance bandwidth definition of the S-parameter (i.e., $|S_{11}| \leq -10$ dB line), the VSWR < 2, reveals the equivalent impedance bandwidth.

The measurements on the VNA also aid in identifying the accurate resonant frequency points in the measured data. Glitches observed in the measurements may be attributed to poor calibration or a noncalibrated VNA. Thus, it is important to calibrate VNA occasionally.

6.4.2 Far-Field Measurement Setup

The far-field radiation patterns are three-dimensional in nature, but they are plotted on two-dimensional scales, namely the φ=0° and the φ=90° plane. These are also referred to as the E-plane pattern (where the E field is dominant) and the H-plane pattern (where the H field is dominant). The most common method of radiation-pattern measurement is the two-antenna method. If the broadest dimension of the DRA is given by D, then the far-field distance of measurement from the source is

$$R = \frac{2D^2}{\lambda} \tag{6.5}$$

where R is the far-field distance between the source (usually a broadband horn antenna) and the AUT, and λ is the operating wavelength corresponding to the measurement frequency.

Once the measurements corresponding to the $\varphi = 0°$ plane are made, the DRA can be rotated by an angle of 90° so as to take the next measurements on the $\varphi = 90°$ plane. The cross-polar components can be taken by a small advantage of either shifting the complete plane of measurements of the AUT. In case the same co-polar and cross-polar components of raditaion is not possible, then the measurements can be done by flipping the broadband-transmitting horn by an angle of 90°. The cross-polar components of the DRA are usually higher than the patch antenna due to the simple boundary conditions that displacement density is discontinuous by the charges stored at the boundary. Also, the complete radiation mechanism of the DRA is governed by the displacement density itself.

The modern antenna positioners are accompanied by software interfaced with the computers that directly map the measured values to the polar plot of the radiation pattern. The same can be then converted into a rectilinear plot, if desired. It is also interesting to note that there are many antenna positioners that have now a servo rotating capability of one-degree measurements. This increases the accuracy of the radiation pattern so when measured and plotted against the simulations, it results in a better and more fair comparison.

Through the radiation pattern, one can easily estimate the 3-dB beamwidth and predict the gain or directivity of the AUT. For example, a pencil beam pattern reflects a very high directivity or gain of the antenna. Similarly, the information obtained from the cross-polar plot along with the copolar plots help estimate a first-level understanding of the polarization of the antenna. If the cross-polar component is 20 dB lower than the copolar component, then the first understanding is that the AUT is likely to be linearly polarized.

6.4.3 Gain and Antenna Radiation Efficiency Measurements

The directivity of an antenna is defined as the ratio of the radiation intensity in a given direction from the antenna to the ratio of the radiation intensity averaged over all directions. The average radiation intensity is equal to the total power radiated by the antenna divided by 4π [39]. This is given as

$$D = \frac{U}{U_0} = \frac{4\pi U}{P_{rad}} \tag{6.6}$$

where D is the directivity (dimensionless), U is the radiation intensity (W/unit solid angle), and P_{rad} is the total radiated power (W).

If the direction is not clearly laid out then the direction of maximum radiation intensity (maximum directivity) is given as

$$D_{\max} = D_0 = \frac{U_{\max}}{U_0} = \frac{4\pi U_{\max}}{P_{rad}} \tag{6.7}$$

where $D_{\max}$ or D_0 is the maximum directivity (dimensionless), and $U_{\max}$ is the maximum radiation intensity (W/unit solid angle).

Similarly, the antenna radiation efficiency is defined as the ratio of the power dissipated into space to the net power delivered to the antenna by the transmitter circuits. It is given as

$$\eta = \frac{R_R}{R_L + R_R} \tag{6.8}$$

where η is the antenna radiation efficiency, R_R is the effective radiation resistance, and R_L is the total loss resistance.

Based on the knowledge of the antenna directivity and the antenna radiation efficiency, the antenna gain can be defined. In a transmitting antenna, the gain describes how well the antenna converts input power into radio waves toward a specific direction. In the case of a receiving antenna, the gain describes how well the antenna converts radio waves arriving toward the antenna from a specified direction to electrical power. Mathematically, the gain is established as

$$G = D\eta \tag{6.9}$$

where the symbols have their usual definitions. Gain is also measured in terms of decibels.

The gain of the antenna can be classically measured using the two-antenna method, which uses the Friis free-space equation:

$$P_R = P_t G_t G_R \left(\frac{\lambda}{4\pi R} \right)^2 \tag{6.10}$$

where P_R is the received power level, P_t is the transmitted power level, G_t is the gain of the transmitting antenna, G_R is the gain of the receiving antenna, and R is the far-field distance between the transmitting broadband horn antenna and the AUT.

As discussed in the measurement setup for the radiation pattern, the VNA is usually connected as the source with 0 to +5 dBm of power. One may also connect a high-power RF source, if desired. Since the gain of the transmitter-calibrated broadband horn antenna is already provided, the received power levels are measured by the plotting software or by means of a spectrum analyzer connected at the far-field distance R operating at a fixed frequency or wavelength λ. Thus, from (6.7), the gain of the receiver or the AUT can be easily calculated at different frequency points and plotted on any plotting tool. In case the broadband horn antenna is not present, then one may fabricate two DRAs with the same physical dimensions and thus assume that $G_t = G_R$, and so the gain of the DRA can be predicted.

From the radiation pattern of both the planes [27], the directivity can be calculated as

$$D = \frac{32400}{\theta_{1D}\theta_{2D}}(HPBW\ in\ degrees) = \frac{4\pi}{\theta_{1D}\theta_{2D}}(HPBW\ in\ radians) \tag{6.11}$$

where θ_{1D} and θ_{2D} are the 3-dB or half-power beamwidths (HPBWs) in the two planes of measurement of the radiation pattern. Once the gain and the directivity of the antenna is known, the antenna radiation efficiency can be calculated from (6.9).

6.4.4 CP Measurements

In EM, polarization refers to restricting the E field component on one direction. This gives rise to three cases: LP, CP, and elliptical polarization. LP means that the E field component is restricted in one axis only. Thus, linearly polarized waves can be either horizontal or vertical or LP at an angle of 45 degrees. If an EM wave is not linear or circular polarized then it is elliptically polarized in nature. In fact, CP is a limiting case of elliptical polarization.

A circularly polarized wave is one where the EM wave is at a constant magnitude and is rotating at a constant rate in a plane perpendicular to the direction of the wave. At any instant of time, the E field vector of the wave indicates a point on the helix oriented along the direction of propagation.

A circularly polarized wave can rotate either clockwise or RHCP or anti-clockwise or LHCP. In RHCP, the E-field vector rotates in the right-hand sense of rotation with respect to the direction of propagation. In LHCP, the E-field vector rotates in a left-hand or counterclockwise sense of rotation with reference to the direction of propagation.

The AR of an antenna is a ratio of the major axis to the minor axis of the rotating E-field vector. Theoretically, for CP, the AR is unity. However, for all practical considerations, the AR in decibels below the -3-dB line is considered as circularly polarized in nature.

The LHCP and RHCP field radiation patterns and the 3-dB AR can be calculated from the measured E-field components of $\overrightarrow{E_\theta}$ and $\overrightarrow{E_\varnothing}$ using (6.12) and (6.13) [40, 41].

$$\overrightarrow{E_R} = \frac{\overrightarrow{E_\theta} + j\overrightarrow{E_\varnothing}}{\sqrt{2}}, \overrightarrow{E_L} = \frac{\overrightarrow{E_\theta} - j\overrightarrow{E_\varnothing}}{\sqrt{2}} \tag{6.12}$$

$$A.R. = \left[\frac{\left|\overrightarrow{E_R}\right| + \left|\overrightarrow{E_L}\right|}{\left|\overrightarrow{E_R}\right| - \left|\overrightarrow{E_L}\right|}\right] \tag{6.13}$$

where $\overrightarrow{E_R}$ is the right-handed E field, and $\overrightarrow{E_L}$ is left-handed E field. Depending on the magnitude of the right-handed and the left-handed field components, LHCP- or RHCP-dominant AUT can be easily established.

6.5 Conclusion

This chapter detailed the practical considerations of the fabrication, testing, and measurement setup generally required for a DRA. The chapter also discussed using Ashby's method to choose a material and describes the experimental machines and setup needed for fabrication, the various fabrication tolerances of the machines, and experimental equipment. The chapter also included references to the equipment manufacturers and material vendors.

References

[1] https://www.rogerscorp.com/.

[2] https://www.laird.com/products/microwave-absorbers/.

[3] https://www.taconic.co.kr/.

[4] Leung, K. W., et al., "Dual-function Radiating Glass for Antennas and Light Covers—Part I: Omnidirectional Glass Dielectric Resonator Antennas," *IEEE Trans. Antennas Propag.*, Vol. 61, No. 2, Feb. 2013, pp. 578-586.

[5] Leung, K. W., et al., "Dual-Function Radiating Glass for Antennas and Light Covers—Part II: Dual-Band Glass Dielectric Resonator Antennas," *IEEE Trans. Antennas Propag.*, Vol. 61, No. 2, Feb. 2013, pp. 587–597.

[6] Chen, Z., H. Wong, and J. Kelly, "A Polarization-Reconfigurable Glass Dielectric Resonator Antenna Using Liquid Metal," *IEEE Transactions on Antennas and Propagation,* Vol. 67, No. 5, May 2019, pp. 3427–3432.

[7] Kumar, J., and N. Gupta, "Investigation on Microwave Dielectric Materials for Dielectric Resonator Antennas," *International Journal of Applied Electromagnetics and Mechanics,* Vol. 47, 2015, pp. 263–272.

[8] Roth, R., F. Field, and J. Clark, "Materials Selection and Multiattribute Utility Analysis," *Journal of Computer-Aided Materials Design,* Vol. 1, 1994, pp. 325–342.

[9] Rao, R. V., "A Material Selection Model Using Graph Theory and Matrix Approach," *Materials Science and Engineering: A,* Vol. 431, 2006, pp. 248–255.

[10] Gupta, N., "Material Selection for Thin-Film Solar Cells Using Multiple Attribute Decision Making Approach," *Materials & Design,* Vol. 32, 2011, pp. 1667–1671.

[11] Reddy, G. P., and N. Gupta, "Material Selection for Microelectronic Heat Sinks: An Application of the Ashby Approach," *Materials & Design,* Vol. 31, 2010, pp. 113–117.

[12] Ashby, M. F., *Materials Selection in Mechanical Design* (Second Ed.), Oxford, United Kingdom: Butterworth-Heinemann, 1999.

[13] Ashby, M. F., et al., "Selection Strategies for Materials and Processes," *Materials & Design,* Vol. 25, 2004, pp. 51–67.

[14] Petosa, A., *Dielectric Resonator Antenna Handbook,* Norwood, MA: Artech House, 2007.

[15] Mukherjee, B., P. Patel, and J. Mukherjee, "A Novel Cup-Shaped Inverted Hemispherical Dielectric Resonator Antenna for Wideband Applications," *IEEE Antenna and Wireless Propagation Letters,* Vol. 12, 2013, pp. 1240–1243.

[16] Hassan, E., E. Wadbro, and M. Berggren, "Patch and Ground Plane Design of Microstrip Antennas by Material Distribution Topology Optimization," *Progress in Electromagnetic Research B,* Vol. 59, 2014, pp. 89–102.

[17] Bhardwaj, A., and V. Dinesh Kumar, "Optical Dot Antenna and Nanohole Transmission," *IET Micro and Nano Letters,* Vol. 7, Issue 11, 2012, pp. 1151–1156.

[18] Kiran, V., S. Dileep, and B. Mukherjee, "Compact Embedded Dual-Element Rectangular Dielectric Resonator Antenna Combining Sierpinski and Minkowski Fractals," *IEEE Transactions on Components, Packaging and Manufacturing Technology,* Vol. 7, No. 5, 2017, pp. 786–791.

[19] Kshirsagar, P., S. Gupta, and B. Mukherjee, "Novel Design of Conformal-Strip Excited Asymmetrical Rectangular Dielectric Resonator Antenna for Ultra-wide Band

Application," *Journal of Microwave Power and Electromagnetic Energy,* Taylor & Francis, Vol. 52, No. 2, 2018, pp. 128–141.

[20] Dileep, S., V. Kiran, and B. Mukherjee, "A Novel Compact Fractal Ring Based Cylindrical Dielectric Resonator Antenna for Ultra-Wideband Applications," *Progress in Electromagnetic Research C (PIER C),* Vol. 67, 2016, pp. 71–83.

[21] Gupta, S., P. Kshirsagar, and B. Mukherjee, "Sierpinski Fractal Inspired Inverted Pyramidal DRA for Wideband Applications," *Electromagnetics,* Taylor & Francis, Vol. 38, Issue 2, 2018, pp. 103–112.

[22] Gupta, S., P. Kshirsagar, and B. Mukherjee, "Low Profile Multilayer Cylindrical Segment Fractal Dielectric Resonator Antenna for Wideband Applications," *IEEE Antenna and Propagation Magazine,* Vol. 61, No. 4, 2019, pp. 55–63.

[23] Rajeshwari, P., et al., "Bandwidth Enhancement for Dielectric Resonator Antenna Using Modified Microstrip Patch Feeding Technique," *International Journal of Engineering Research and Technology (IJERT) NCEICC,* Vol. 5, Issue 9, 2017, pp. 1–3.

[24] Zubir, I. A., "A LOW PROFILE Hybrid Multi-Permittivity Dielectric Resonator Antenna with Perforated Structure for Ku and K Band Applications," *IEEE Access,* Vol. 8, 2020, pp. 151219–151228.

[25] Dileep, S., V. Kiran, and B. Mukherjee, "Compact Bi-Cone Dielectric Resonator Antenna for Ultra-Wideband Applications—A Novel Geometry Explored," *Electromagnetics,* Taylor & Francis, Vol. 37, Issue 7, 2017, pp. 471–481.

[26] Mukherjee, B., et al., "A Novel half Hemispherical Dielectric Resonator Antenna with Array of Slots for Wideband Applications," *Progress in Electromagnetic Research C (PIER-C),* Vol. 36, 2013, pp. 207–221.

[27] Mukherjee, B., P. Patel, and J. Mukherjee, "Hemispherical Dielectric Resonator Antenna Based on Apollonian Gasket of Circles—A Fractal Approach," *IEEE Transactions on Antennas and Propagation,* Vol. 62, No. 1, 2014, pp. 40–47.

[28] Gupta, S., et al., "Compact and Circularly Polarized Hemispherical DRA for C-Band Applications," *Frequenz- Journal of RF Engineering and Telecommunications,* Vol. 73, Issue 7–8, 2019, pp. 227–234.

[29] Ali, S. M., et al., "Recent Advances in Wearable Antennas in Materials, Fabrication Methods, Designs and Their Applications: State of the Art," *MDPI Micromachines,* Vol. 8, 2020, pp. 1–41.

[30] Xia, Z. X., K. W. Leung, and K. Lu, "3D-Printed Wideband Multi-Ring Dielectric Resonator Antenna," *IEEE Antenna and Wireless Propagation Letters,* Vol. 18, No. 10, 2019, pp. 2110–2114.

[31] Nayeri, P., et al., "3D printed Dielectric Reflect Arrays: Low-Cost High-Gain Antennas at Submillimeter Waves," *IEEE Transactions on Antennas and Propagation,* Vol. 62, No. 4, 2014, pp. 2000–2008.

[32] Addamo, G., et al., "3D printing of High-Performance Feed Horns from Ku to V Bands," *IEEE Antenna and Wireless Propagation Letters,* Vol. 17, No. 11, 2018, pp. 2036–2040.

[33] Nayeri, P., and G. Brennecka, "Wideband 3D Printed Dielectric Resonator Antennas," *Proceedings of APS/URSI UNSC Meeting,* 2018, pp. 2081–2082.

[34] Huang, J., et al., "Impact of infill Pattern on 3D Printed Dielectric Resonator Antennas," *IEEE Asia Pacific Conference on Antennas and Propagation (APCAP),* Auckland, 2018, pp. 233–235.

[35] Basu, A., *An Introduction to Microwave Measurements,* Boca Raton, FL: CRC Press, 2017.

[36] https://www.keysight.com/.

[37] https://www.rohde-schwarz.com/.

[38] https://www.anritsu.com/.

[39] Balanis, C. A., *Antenna Theory: Analysis and Design,* John Wiley and Sons, 2005.

[40] Chauhan, M., A. K. Pandey, and B. Mukherjee, "A Novel Cylindrical Dielectric Resonator Antenna Based on Fibonacci Series Approach," *Microwave and Optical Technology Letters,* Wiley, Vol. 61, No. 10, 2019, pp. 2268–2274.

[41] Yang, S. I. S., et al., "Study on Sequential Feeding Networks for Subarrays of Circularly Polarized Elliptical Dielectric Resonator Antenna," *IEEE Transactions on Antennas and Propagation,* Vol. 55, No. 23, Feb. 2007, pp. 321–33.

About the Authors

Biswajeet Mukherjee received his bachelor's degree in Electronics and Communication Engineering (ECE) from Indraprastha University, Delhi, India; his master's degree in microwave electronics from the Department of Electronic Science, University of Delhi, India; and his Ph.D. from the Indian Institute of Technology (IIT) Bombay, India. He is currently working as an assistant professor in the Department of Electronics and Communication Engineering, IITDM, Jabalpur, India. He has authored or coauthored more than 90 research articles in peer-reviewed journals and conferences. His current research interests include dielectric resonators, antennas, MICs, MMICs, and computational electromagnetics in FDTD. He is a Senior Member of the IEEE, a senior member of URSI, a member of the Institution of Engineers India (IEI), and a member of EuRAAP. He has supervised two Ph.D. students and currently has four students completing their doctorates under him. Further, he has guided eight master's students. Mukherjee is the principal investigator of six sponsored projects from agencies of the Indian government, including the Department of Science and Technology and the Ministry of Science and Technology. He received the "Best Paper Award" at the International Conference on Innovative Product Design and Intelligent Manufacturing Systems (ICIPDIMS), 2019, in NIT Rourkela, India. In addition, he was awarded the Hari Om Ashram Prerit Harivallabhdas Chunilal Shah Research Endowment Prize in the field of electronics and communication for the year 2015–16, declared in 2017, by Sardar Patel University, Gujarat. He is also a recipient of the Indian National Science Academy Visiting Scientist Fellowship for 2017–18 and was awarded the IEI's "Young Engineer Award" for his contribution to the field of electronics and telecommunication, in Vijaywada, India, 2016. Moreover, he received the "Young Scientist Award" at the 31st URSI General Assembly and Scientific Symposium (GASS) at Beijing, China, August 16–23, 2014.

Monika Chauhan received her bachelor's of engineering degree in electronics and telecommunication from the Institue of Engineering and Technology (IET), Devi Ahilya Vishwavidyalaya (DAVV) University, Indore, India, and her master's in engineering in microwaves from the Department of Electronics and Communication at Jabalpur Engineering College (JEC), Rajiv Gandhi Proudyogiki Vishwavidyalaya (RGPV) University, Jabalpur, India. Dr. Chauhan completed her Ph.D. degree in the Department of Electronics and Communication Engineering, IIITDM, Jabalpur, India. Her research areas of interest include planar antennas, dielectric resonator antennas, metamaterials, and filters. She is a recipient of the "Young Scientist Award" from the Madhya Pradesh Council of Science and Technology.

Index

Recent Titles in the Artech House Antennas and Propagation Library

Christos Christodoulou, Series Editor

Adaptive Array Measurements in Communications, M. A. Halim

Antenna-on-Chip: Design, Challenges, and Opportunities, Hammad M. Cheema, Fatima Khalid, and Atif Shamim

Antenna Design for Cognitive Radio, Youssef Tawk, Joseph Costantine, and Christos Christodoulou

Antenna Design with Fiber Optics, A. Kumar

Antenna Engineering Using Physical Optics: Practical CAD Techniques and Software, Leo Diaz and Thomas Milligan

Antenna Measurement Techniques, Gary E. Evans.

Antennas and Propagation for Body-Centric Wireless Communications, Second Edition, Peter S. Hall and Yang Hao, editors

Antennas and Site Engineering for Mobile Radio Networks, Bruno Delorme

Antennas for Small Mobile Terminals, Kyohei Fujimoto and Koichi Ito, editors

Analysis, Design and Measurement of Small and Low-profile Antennas, Kazuhiro Hirasawa

Analysis of Radome-Enclosed Antennas, Second Edition, Dennis J. Kozakoff

Applications of Geographic Information Systems for Wireless Network Planning, Francisco Saez de Adana, Josefa Gómez Pérez, Abdelhamid Tayebi Tayebi, and Juan Casado Ballesteros

The Art and Science of Ultrawideband Antennas, Second Edition, Hans G. Schantz

AWAS for Windows Version 2.0: Analysis of Wire Antennas and Scatterers, Antonije R. Djordjević, et al.

Broadband Microstrip Antennas, Girsh Kumar and K. P. Ray

Broadband Patch Antennas, Jean-François Zürcher and Fred E. Gardiol

Circularly Polarized Dielectric Resonator Antennas, Raghvendra Kumar Chaudhary, Rajkishor Kumar, and Rakesh Chowdhury

Design and Applications of Active Integrated Antennas, Mohammad S. Sharawi and Oualid Hammi

Designer Notes for Microwave Antennas, Richard C. Johnson

Dielectric Resonator Antenna Handbook, Aldo Petosa

Dielectric Resonator Antennas, Biswajeet Mukherjee and Monika Chauhan

Electromagnetics and Antenna Technology, Alan J. Fenn

Electromagnetics, Microwave Circuit and Antenna Design for Communications Engineering, Peter Russer

Engineering Applications of the Modulated Scatterer Technique, Jean-Charles Bolomey and Fred E. Gardiol

Fixed & Mobile Terminal Antennas, Akhileswar Kumar

Foundations of Antenna Engineering: A Unified Approach for Line-of -Sight and Multipath, Per-Simon Kildal

Four-Arm Spiral Antennas, Joseph A. Mosko

Frequency-Agile Antennas for Wireless Communications, Aldo Petosa

Fresnel Zones in Wireless Links, Zone Plate Lenses and Antennas, Hristo D. Hristov

Handbook of Antennas for EMC, Second Edition, Thereza Macnamara with John McAuley

Handbook of Reflector Antennas and Feed Systems, Volume I: Theory and Design of Reflectors, Satish Sharma, Sudhakar Rao, and Lotfollah Shafai, editors

Handbook of Reflector Antennas and Feed Systems, Volume II: Feed Systems, Lotfollah Shafai, Satish Sharma, and Sudhakar Rao, editors

Handbook of Reflector Antennas and Feed Systems, Volume III: Applications of Reflectors, Sudhakar Rao, Lotfollah Shafai, and Satish Sharma, editors

Interference Suppression Techniques for Microwave Antennas and Transmitters, Ernest Freeman

Introduction to Antenna Analysis Using EM Simulators, Hiroaki Kogure, Yoshie Kogure, and James C. Rautio

LONRS: Low-Noise Receiving Systems Performance and Analysis Toolkit, Charles T. Stelzried, Macgregor S. Reid, and Arthur J. Freiley

Measurement of Mobile Antenna Systems, Second Edition, Hiroyuki Arai

Microstrip Lines and Slotlines, Third Edition, Ramesh Garg Inder J. Bahl, Maurizio Bozzi

Microstrip Antenna Design, K.C Gupta and Abdelaziz Benalla

Microwave and Millimeter-Wave Remote Sensing for Security Applications, Jeffrey A. Nanzer

Microwave Cavity Antennas, Akhileswar Kumar and Hristo D. Hristov

Millimeter Wave Microstrip and Printed Circuit Antennas, Prakash Bhartia

Mobile Antenna Systems Handbook, Third Edition, Kyohei Fujimoto, editor

Moment Methods in Antennas and Scattering, Robert C. Hansen

Multiband Integrated Antennas for 4G Terminals, David A. Sánchez-Hernández, editor

Near-Field Antenna Measurements, Dan Slater

Noise Temperature Theory and Applications for Deep Space Communications Antenna Systems, Tom Y. Otoshi

Phased Array Antenna Handbook, Third Edition, Robert J. Mailloux

Phased Array Antennas, Arthur A. Oliner

Phased Array Antennas with Optimized Element Patterns, Sergei P. Skobelev

Plasma Antennas, Second Edition, Theodore Anderson

Polarization in Electromagnetic Systems, Second Edition, Warren Stutzman

Practical Antenna Design for Wireless Products. Henry Lau

Practical Microstrip and Printed Antenna Design, Anil Pandey

Practical Phased Array Antenna Systems, Eli Brookner

Practical Simulation of Radar Antennas and Radomes, Herbert L. Hirsch

Printed MIMO Antenna Engineering, Mohammad S. Sharawi

Radiowave Propagation and Antennas for Personal Communications, Third Edition, Kazimierz Siwiak

Reconfigurable Antenna Design and Analysis, Mohammod Ali

Reflectarray Antennas: Analysis, Design, Fabrication and Measurement, Jafar Shaker, Mohammad Reza Chaharmir, and Jonathan Ethier

Reflector Lens Antennas: Analysis and Design Using Personal Computers, Carlyle J. Sletten, editor

RF Coaxial Slot Radiators: Modeling, Measurements, and Applications, Kok Yeow You

Shipboard Antennas, Second Edition, Preston Law

Significant Phased Array Papers, Robert C. Hansen

Smart Antenna Engineering, Ahmed El Zooghby

Solid Dielectric Horn Antennas, Carlos Salema, Carlos Fernandes, and Rama Kant Jha

Ultrawideband Antennas for Microwave Imaging Systems, Tayeb A. Denidni and Gijo Augustin

Ultrawideband Short-Pulse Radio Systems, V. I. Koshelev, Yu. I. Buyanov, and V. P. Belichenko

Waveguide Components for Antenna Feed Systems: Theory and CAD, Jaroslaw Uher, Jens Bornemann, and Uwe Rosenberg

For further information on these and other Artech House titles, including previously considered out-of-print books now available through our In-Print-Forever® (IPF®) program, contact:

Artech House
685 Canton Street
Norwood, MA 02062
Phone: 781-769-9750
Fax: 781-769-6334
e-mail: artech@artechhouse.com

Artech House
16 Sussex Street
London SW1V HRW UK
Phone: +44 (0)20 7596-8750
Fax: +44 (0)20 7630 0166
e-mail: artech-uk@artechhouse.com

Find us on the World Wide Web at: www.artechhouse.com